BMW MOTORCYCLES

Gold Portfolio

1971-1976

Compiled by
R.M.Clarke

ISBN 1 85520 3782

Brooklands Books

BROOKLANDS BOOKS LTD.
P.O. BOX 146, COBHAM,
SURREY, KT11 1LG. UK

A -BM71BGP

Printed in Hong Kong

Brooklands Books

ACKNOWLEDGEMENTS

To have everything of any significance about a particular make and model of motorcycle within a single book is a decided advantage. This was Booklands Books objective when they launched their Gold Portfolio series in the knowledge that it would save enthusiasts having to trace an elusive feature by wading through a pile of back issues of old magazines.

This, the second of two BMW titles, relates solely to the twin cylinder range manufactured from 1971 onwards. It covers the time when the /5 series represented an entirely new generation of machines to keep BMW abreast of current design trends. It also coincided with the beginnings of the superbike era, into which the larger capacity models fitted.

Without the generous assistance given by the world's leading publishers, this collation of comprehensive information would not have been possible. Brooklands Books are very much indebted to the publishers of *Bike, Cycle, Cycle Buyers Guide, Cycle Guide, Cycle Rider, Cycle World, Motorcycle Mechanics, Motorcycle Sport, Motorcyclist Illustrated, Motorcyclist Illustrated Annual* and *Two Wheels* for permission to reproduce their copyright material within the covers of this book.

R.M. Clarke

BMW has a wealth of experience in producing motorcycles with a horizontally-opposed twin cylinder engine and shaft final drive. Indeed, it is these features that characterise the BMW, allied with the high quality of their product and the exceptionally low level of mechanical noise. Never cheap in comparison with their contemporaries, they attracted the more discerning motorcyclist to whom price was a secondary consideration.

The /5 series represented a determined attempt by BMW to have an entirely up-to-date model range in capacities ranging from 498 to 898cc's. Now the motorcycles were being made in Spandau as car manufacture had taken up all the production space in Munich. The /5 series duplex tube frame was reminiscent in some respects of the Norton 'featherbed' design, having a heavily braced steering head but only a single larger diameter top tube with a secondary lower tank rail. The rear sub-frame was a bolted-on assembly, its right-hand tube enclosing the shaft final drive. A conventional telescopic front fork now replaced the earlier Earles design and the new engine unit was massive, virtually filling the frame. Tilted slightly upwards at the front, it provided better angular travel for the universal joint of the shaft final drive. It also now contained an electric starter, regarded as an essential part of every modern motorcycle. For the first time BMW had available alternative colour schemes in addition to the traditional all black finish with white lining.

The /6 series that followed had further refinements which included a five-speed gearbox and an hydraulic disc front brake. New long, tapered silencers ensured noise levels were kept to a minimum with increasing concern being shown about the environment. BMW were very much back in the running and were being used by many police forces for law enforcement on account of their reliability.

Jeff Clew

Brooklands Books

CONTENTS

MOTORCYCLE
SCOOTER & THREE-WHEELER
MECHANICS

ROAD TEST

THE

Two wheels versus four . . . with the police as official observers on a timed run from London to Birmingham and back! The BMW 750 scored a convincing success over a Lotus Elan as the best means of transport through the city traffic . . . all this and more from the latest bike built in Berlin, Germany . . .

It doesn't matter how much money a manufacturer spends on advertisements that sing the praise of his wares; if he stays in business long enough, a reputation—be it for good or ill—will be earned.

Some love Be-ems, others hate 'em, but nobody can deny that they are beautifully engineered machines with a name which has become synonymous with two-wheeled excellence.

How do the factory achieve such good results? *MM* staff were recently invited to the factory in West Berlin to see. The answer is stringent quality control, both mechanical and visual. Every engine is bench-tested and every bike is track-tested—not just a belt round the block, but thoroughly tested—before it leaves the BMW works.

It is interesting to note that the lines on the mudguards are painted by hand. They say this is the best way. Anyhow, the end product is one of the best finished bikes you can buy, so the man who said "You get what you pay for!" probably had something.

STARTING

Starting the R75/5 was never a problem. On cold mornings, out of sympathy for the battery, the motor was turned over by kickstart a couple of times to free it off. The air lever was then half closed, ignition turned on and the starter button would instantly bring the motor to life. After a few hundred yards, the air lever could be fully opened and the motor would run and idle without fuss.

Unfortunately, on the bike we had, there was a bad contact on the starter button and some-

LOTUS EATER!

In the headlamp is the speedometer/rev counter unit. The ignition switch is traditional Be-em type. Below is the neat shaft drive which remained oiltight throughout test

times it took several presses to activate the starter. When the kickstarter was used, it was simple to operate (despite its unorthodox action) and was always a first-timer.

Acceleration was deceptively good. The quietness of the engine and the lack of wind rush, due to the fairing, gave one the impression that the bike was going more slowly than it was.

BMW claim the top speed to be about 110mph. But their aim is not to make a super-fast machine. The R75/5 is in fact greatly de-tuned, so that it is a fast, safe, touring bike. The carburettors, for instance, are constant velocity type for this very reason.

When ordinary carburettors were fitted, it was found that too much power was produced low down. The result was wheel-spin in first and second gears. The constant velocity carb ensures that the perfect mixture is supplied at all throttle openings. Power now comes in smoothly low down and is completely controllable.

The R75/5 was obviously built with autobahns in mind, and not unnaturally, this is where the bike comes into its own. Cruising at 90–95 mph could be held indefinitely. It was only the speed with which one caught and passed other road users that gave any sensation of speed at all.

The riding position was very comfortable and no fatigue was felt even after 100 miles non-stop riding. The only discomfort felt was caused by the shins knocking against the carburettor tops.

ROADHOLDING

Roadholding was excellent and the suspension, with its exceptionally long travel, $8\frac{1}{2}$-inch front and 5-inch rear, ironed out all normal road bumps. It steered well and stuck firmly on line, providing the power was kept on.

The flat twin layout makes itself felt by a slight torque reaction when power is suddenly shut off. Do this into a bend and things get a bit wavery. The steering damper was not used at all throughout the test.

Fairings, as all good motorcyclists know, magnify engine noise, sometimes to an alarming degree. The Be-em is exceptionally quiet, both at engine and exhaust pipe ends. German noise regulations are strict!

There was, however, a whine in first, second and third gears, but top was absolutely quiet. It is only fair to say that without the fairing, the noise would probably have gone unnoticed.

The clutch is of course run at engine speed, which brings us to the biggest fault with flat twins—the gearchange. The gears themselves were well spaced, if rather wide between bottom and second. Even in dense traffic, where big, heavy bikes are not always at their best, there was no need for too much cog-swapping.

The gearchange was, like all other flat-twin gearchanges, clonky. Heavyish flywheels and engine-speed clutch are not the most conducive to smooth, silent changes and this was the feature which stood out like a sore thumb on an otherwise first-class machine.

There were all sorts of nadgery techniques which would sometimes bring about quietish changes, but why should one have to resort to this sort of

ROAD & TRACK TEST

BMW R75/5

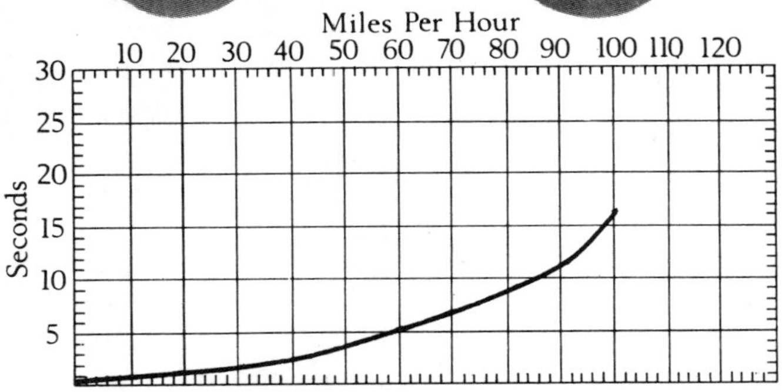

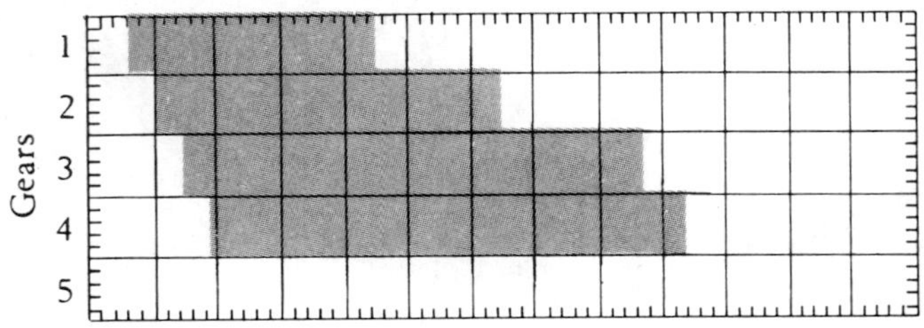

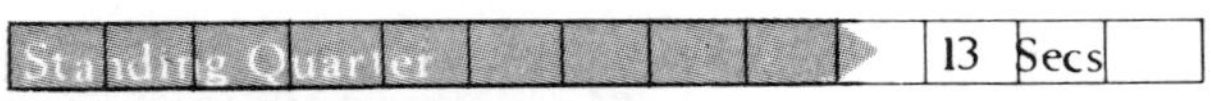

fuel consumption (average figure) **50 mpg**
braking from 30 mph **29 ft.**

SPECIFICATION

Engine: Twin-cylinder, horizontally opposed, overhead valve with air cooling. Bore 82 mm, stroke 70.6 mm, compression ratio 9:1. Lubrication by wet sump, oil capacity four pints. Carburation by two Bing constant-velocity, 32 mm choke carbs. Paper element air cleaners. Claimed bhp 50 at 6200.

Transmission: Single-plate dry clutch with diaphragm spring on crankshaft. Final drive by enclosed shaft and helical bevel gears. Gear ratios: 11.34, 7.5, 5.46, 4.37 to 1. 30 mph in top gear equals 1700 rpm. Gears: foot-change operated, left-hand side.

Electrics: 12-volt battery and coils. Bosch alternator and flashing indicators.

Tyres: Metzeler, 3.25 × 19 front, ribbed. 4.00 × 18 studded rear. Brakes: 8-inch tls front and 8-inch sls rear.

Suspension: Telescopic front forks with two-way damping. Swinging arm rear, with hand-adjustable dampers.

Dimensions: Wheelbase 54½ in., ground clearance (unladen) 7 in. Dry weight 419 lb.

Price: £1121·40 including British purchase tax.

Importers: BMW Concessionaires (GB) Ltd, 361/365 Chiswick High Road, London, W.4.

Bavarian Motor Works . . .

thing on a high-quality machine?

The clutch showed no signs of slipping, even in the densest of traffic, although we were warned that constant slipping in heavy traffic, etc, might produce heat distortion after a while. It was quite a pleasant clutch to use; not too heavy and, despite its bulk, smooth in take-up.

Brakes were very good. The front brake, an 8-inch twin-leading-shoe, was extremely powerful and showed no sign of fade from ton-plus stops. On greasy roads, care was needed when applying the front brake though. It had a tendency to grab, possibly due to its newness, but anyway it was treated with extreme caution in the wet.

The rear brake, an 8-inch single-leading-shoe, was also powerful and together the brakes recorded some very respectable stopping distances.

LIGHTING

Motorcycle electrics have improved greatly over the last decade and the lights fitted to

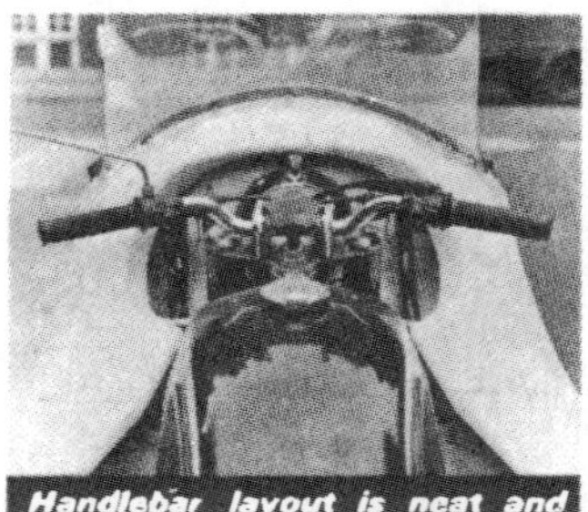

Handlebar layout is neat and controls well placed. The steering damper was not used at all

this Be-em were excellent. Cruising at the legal limit was completely safe and the dip-switch was such that there was no fear of upsetting oncoming drivers.

The dipper also serves to flash the main beam and proved very useful for overtaking at night. The horn was audible to the rider, but hardly so to anyone else. With a machine of this power, a really powerful horn should be fitted, especially as the exhaust note is so unobtrusive.

Our only trouble came, when a filament blew in the rear light. You can't change the snap connectors over to stop-light—there aren't any on the Be-em, so as a get-you-home dodge, out came the piece of silver paper to bridge the bulb contacts. Unfortunately, although this gave a light, it also made the footbrake into another flasher! Every time the footbrake was applied, the headlamp came on! Weird things, these German electrics!

Mounted in the headlamp is the speedo/tachometer combined unit. The speedometer was 5 mph optimistic at 90 mph and the tacho was partially obscured by

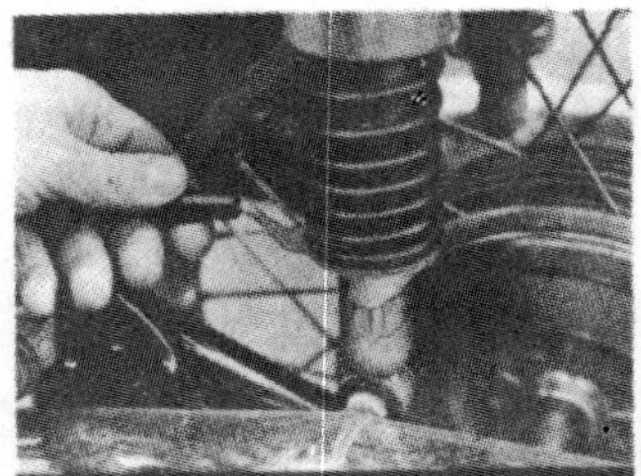

Rear suspension dampers can be adjusted by hand in seconds to suit conditions and weight

the clutch and brake control cables. The ignition switch is the traditional BMW push-down type with cover which slides back to keep out the elements.

The toolbox, situated beneath the dualseat, is lockable and the kit inside was comprehensive by modern standards. Flashing indicators are the rule rather than the exception nowadays and these were mounted neatly in the fairing and operated by a switch next to the throttle grip and starter button.

All controls were well placed and the levers had finger adjustments to take the odd slack from cables *en route*.

OILTIGHT

As one would expect, the motor remained perfectly oil-tight throughout the whole test, as did the transmission. The only evidence that the bike had been ridden was the film of road dust and flies spattered over the headlamp glass, plus the blued-up exhaust pipes—they all do it!

As for the screen, this was an Avon and is available in various colours from £46 to £50. This undoubtedly contributed to the fatigue-free riding which we enjoyed on the bike.

To sum up the Be-em, what can one say? It is fast, reasonably economical and has a beautiful finish. Mechanically it must be one of the world's finest ever two-wheelers.

Undoubtedly it is an enthusiast's machine—who else would pay over £1000 for a motorcycle? Nevertheless, if you have the cash and you want the best—and can afford the spares for it—the R75/5 could be the bike!

BMW R 75/5

CYCLE ROAD TEST:

BMW 750cc R 75/5

Neither a superbike nor a pure touring mount, the BMW 750 occupies a middleground. Slower than some superbikes point-to-point, the BMW is also more fuss-free. The R 75/5 lacks the armchair comfort of big touring rigs—and their truckiness.

Everybody knows something about BMWs. Especially the good old BMWs of the mid-1960s. About five years ago, back when other big machines could rattle sewer grates with exhaust racket, the German twins moved in near-silence. At a time when other big-inchers sprayed parts and numbed riders with old Down-The-Road Palsies, BMWs neither shook nor sprayed. Nor leaked. While some bikes stopped on the highway with great noise and conviction, the black twins still hummed on for miles. At nightfall other headlamps gave off cold and indifferent glows, but BMW washed the road in a flood of light. Back then, the nice part about a BMW trip was that—upon departure—you knew you'd travel in comfort. And in confidence. Because you would arrive. That was the old BMW, and that was 1965.

Today is 1971. Reliable? So what? A whole generation of new machines are trustworthy across long stretches. Comfortable? Several contemporary big bikes won't raise saddlesores inside 250 miles. Smooth? So are today's threes and fours—and some twins. Quiet? Mufflers which modern bikes wear let the sewer grates lie in peace. Lights? Even the Italians have gotten the message. Shaftdrive? No longer a BMW exclusive in the American market. Dripfree? Lots of machines can hold their oil now. One-kick starting? No big deal; electric starters have arrived. And big bikes in 1971 handle and brake—and they top it off with horsepower, barnsful of the stuff. The old BMW was a great machine for its day; and if made today, it would still be a good one. But the new BMW, specifically the R 75/5, is a better machine.

The R 75 is also the best BMW in the revamped series of opposed twins. This shouldn't be surprising. BMW designed one motorcycle, but built three different models. The Germans drew the blueprints with the 750cc engine in mind and then plugged in the necessary components to make the same motorcycle in 500cc and 600cc versions. The "smaller" models in such an arrangement usually suffer. In its road test of the R 60/5 (Nov. 1970), *Cycle* quarreled with the power delivery of the 600 version. In buying performance and paying for this by narrowing the powerband, BMW flawed the R 60's touring charm. The 750cc is no victim of such a trade-off. The R 75 has horses everywhere, and its appearance—awesome in black—suggests the power and pull of a giant locomotive.

Substance, the R 75 has it. Approach the machine; you can't avoid the substance. The mass and density of the thing confronts you everywhere. Viewed from

PHOTOGRAPHY BY DAVID GOOLEY

the side, the R 75 presents closed, hard, smooth surfaces to the eye, with the tank and engine case dominating the bulwark. No large, airy spaces anywhere. It's the kind of machine that reaches out and seizes all the space which it unyieldingly occupies.

Black paint fits the mood the machine creates. A white or silver tank sitting atop the engine slices the motorcycle into thirds by erecting a vertically oriented plane right in the center of the machine. On the other hand, a black tank and fenders make a rather boxy machine look all of one piece, by connecting the sections of the bike. The big black tank counterpoints the engine and holds it in check. The paintwork itself is first-quality everywhere. Other manufacturers have come a long way in the paint department, but no one has matched the BMW black for depth and luster. And the durability of the paint is a matter of record. A black BMW just looks right. It's in keeping with the substantive character of the machine.

When you drop into the saddle, you keep getting that feeling of substance and unbreakability. The major pieces look like they could withstand blows from a Number Nine Hammer. The handlebar switches—the lefthand for horn, high/low beam and headlight flasher; the righthand for the turn signals and starter—move stop to stop with satisfying determination. There are no vague in-between positions. The flip-up gas cap opens with a positive click. Everything knows its place. Even the rider's fingers get organized, because the handlevers have comfortable finger scallops.

Up front on the headlight nacelle the speedometer and tachometer share a cavity, some $3^1/_2$ inches in diameter. The dial sweep of the speedo is about 200 degrees; the smaller tach wears a 270 degree face. The white numbers and needles stand out on black backgrounds, but the numbers are less than $^3/_{16}$ths of an inch high. Cables don't cross the faces so the instruments can be read. Getting a read on speed and rpm would be easier, especially at night, if BMW had designed separate instruments and mounted them outside the headlight nacelle. The speedometer quarters four indicator/warning lights: high beam, neutral, oil pressure, and alternator. In the press for space, the BMW trip odometer has been bumped out completely. New for 1971 is a fifth light. The turn signal indicator light is positioned in the front left quarter of the nacelle. Since a great number of BMW owners outfit their machines with windshields and handlebar fairings, which block the rider's view of the directional signals, the winker light makes sense—until some wise man devises self-cancelling motorcycle turn signals.

The old-time plunger switch which controls ignition, headlight and parking light is still around on the new BMWs. The plunger has a lot of things going for it: it's easy to find in your pocket, even with gloves; it snaps down into the "off" position and won't jiggle out; it won't scarf up the paint, and the plunger receptacle has a spring loaded cover which keeps the wet out when the plunger key is withdrawn. Nevertheless, the universal plunger key doesn't offer the theft protection which a normal key does. Had BMW sunk as much thought into developing a locking key switch as they have put into the old plunger arrangement, BMW could have created a superior switch utilizing a normal key. And then added an emergency off-switch adjacent to the twistgrip.

Two things do key-lock; the forks and the push button latch on the saddle. Once opened, the saddle uncovers a tool tray, in which the rider finds a handcloth and a kit with good quality tools. A tire pump attaches on the inside of the upper left frame member. There are enough tools to take any owner as far as he ought to go inside the machine. The tool tray, nestled between frame tubes, can be pulled out for servicing the battery. Removing the tank only requires spinning off two wing-nuts at the rear of the tank, disconnecting the fuel lines, and lifting the six-and-some-odd gallon reservoir off its rubber cushions. This exposes the ignition coils, regulator, and starter relay. The ease with which these components can be reached makes life more pleasant for BMW mechanics and service-conscience owners.

Pull in the sidestand, open the fuel taps, push the ignition plunger down, watch the indicator warning lights go on, hit the starter button, and listen to the engine settle down to an 800 rpm idle. Draw the clutch lever; drop the transmission into first. The clutch lever pulls lightly enough, and the clutch feeds in the power quickly without the first hint of slip. The torque reaction, produced by the directional plane in which the opposed twin runs, is evident at first. Then, as the mind adjusts to it, the left-to-right rocking upon take-off fades out of mainstream awareness.

Out on the road no one will ever mistake the 1971 R-75 for a 1965 majestic-but-slow Munich twin. In fact, the 1971 R 75 is snappier than the 1970 R 75, and the difference is in the gearing. This year's 750 pulls lower gearing (higher numerically). BMW regeared the rear-end, switching from a 11/32 to a 10/32 combination. In 1970 fourth gear was just too tall when R 75s pushed around a windshield and carried loaded saddlebags. And the overgearing took the bite out of the powerful engine. With 1971 gearing and the 750cc engine, the BMW never wheezes along in some rpm range where it has little pep. The machine, quite unlike *Cycle*'s test R 60, has muscle everywhere. Fourth-gear passing on the open road is effortless; just a crack of the wrist produces the action. If you like to sweep by slow-moving objects with a vindicative rush, catch third, blast the dawdler with the horn, and redline the tach.

Carb air intake tubes can foul legs.

Tools and pump live under hinged saddle.

Plunger operates ignition and lights.

The
BMW R 75/5,
smooth,
silent and
strong,
loves both
the super-
pikes
which break
the land-
scape and
those
two-lane
byroads
long
forgotten.

'he 750 mounts constant vacuum carbs.

The engine sends out some vibrations below 3500 rpm; the vibes reach the rider through the handlebars, and to a lesser extent, through the tank and pegs. Noticable, but not annoying. Above 3500 rpm things really smooth out. Our test machine had a slight trace of vibration in fourth gear at 75 mph. Just over 70 mph the R 75/5 proved incredibly smooth—nary a shake no place. And if one observes Nevada speed limits (in Nevada or elsewhere), the R 75 flies just as smoothly at 90 mph as it does at 70 mph.

The 32mm Bing constant velocity carburetors produce an even power delivery by metering in only as much fuel and air as the cylinders can draw. It's not possible to gag the engine by suddenly wacking the throttle full-on. The Bing CV units have detailed changes for 1971. The flexible neoprene diaphram which attaches to the top of the carb slide and separates the pressure masses is now constructed of heavier material. Earlier (and thinner) diaphrams could be ruptured if not handled carefully. And the carb slides run with greater slide-to-wall clearance in the 1971 models. Greater clearance prevents galling and sticking slides. The new carburetors got blamed at first for an idling problem which the vergassers had nothing at all to do with. When cold the R 75 would idle perfectly; but when hot, the idling speed climbed to 1800 rpm and showed no inclination to drop back down. A quick trip to the Flanders Company (Western BMW distributors) brought an instant diagnosis and solution: the automatic advance unit had been hanging up. Fifteen minutes of fast, accurate work set things right—and the hot idle stayed below a 1000 rpm.

The front brake would haul the R 75 down quickly, evenly, and in a straight line. The double leading shoe front brake is absolutely first rate. The front stopper had linings which were well bedded-in; braking was much better than *Cycle*'s 1970 test R 60. Progressive and predictable, the R 75's front brake feeds the rider clear signs through the handlever as the outer limit of adhesion approaches. The way in which the twin brake-arms on the front backing plate connect together seems a bit curious. Normal practice joins the two brake arms with an adjustable rod, and both arms pull in the same direction. BMW does things differently. The front brake cable attaches directly to both arms in such a way that the cable draws the arms together. The BMW method would appear to be more difficult to adjust precisely; but properly set, there's no doubt that the BMW front stopper really works well. Good thing, because the single leading shoe rear brake needed some attention: it was a squeaker which hadn't bedded properly, and consequently pulled down roughly and snatched a little.

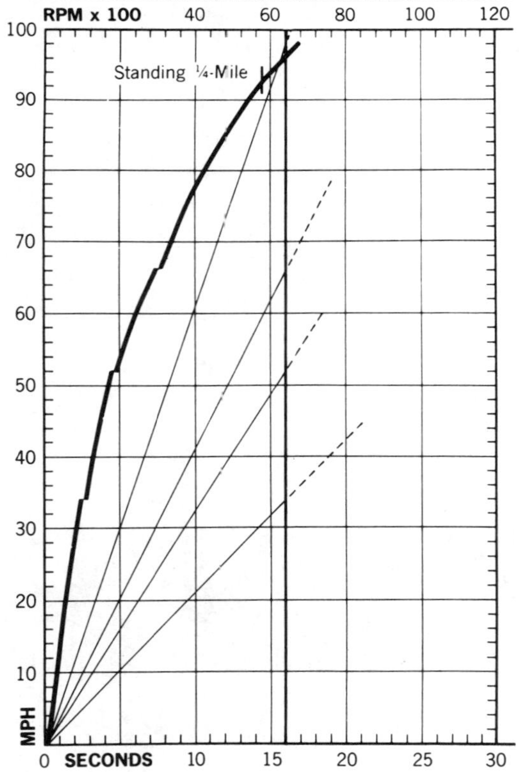

BMW R 75/5 750

Price, suggested retail	USA, POE $1848
Tire, front	3.25 in. x 19 in.
rear	4.00 in. x 18 in.
Brake, front	7.9 in. x 1.2 in.
rear	7.9 in. x 1.2 in.
Brake swept area	59.52 sq. in.
Specific brake loading	9.67 lb/sq. in., at test weight
Engine type	OHV opposed Twin
Bore and stroke	3.22 in. x 2.77 in., 82mm x 70.6mm
Piston displacement	45.46 cu. in., 745cc
Compression ratio	9:1
Carburetion	2; 32mm; Bing vacuum
Air filtration	Dry paper
Ignition	Battery and coil
Bhp @ rpm	57 @ 6400 rpm
Mph/1000 rpm, top gear	15.33
Fuel capacity	6.3 gal.
Oil capacity	4.75 pints
Lighting	12v, 180 watts
Battery	12v, 15ah
Gear ratios, overall	(1) 13.34 (2) 8.72 (3) 6.20 (4) 4.92 (5) none
Wheelbase	54.5 in.
Seat height	32 in., with rider
Ground clearance	6.5 in., with rider
Curb weight	440 lbs., with ½-tank of gas
Test weight	615 lbs., with rider
Instruments	120 mph Speedometer 7500 rpm Tachometer
0-60 mph	6.2 seconds
Standing start ¼ mile	14.48 seconds 92.97 mph
Top speed	N.A. mph

BMW R 75/5 *Continued*

Most machines which weigh in arouna 450 pounds feel trucky around town. The R 75 is amazing. The rider needn't wrestle it in downtown traffic. The new BMW has less to toss around than the old twins; the R 75 weighs a good 40 pounds less than its predecessor, the R 69S. Pounds have come off in several places. The fenders for example, are fiberglass. And the bolt-together frame is lighter than the old one-piece models. The rear section simply bolts to the main double loop frame. The bolt-up construction is handier and less expensive to ship and straighten in the event of accident damage. Stare at those big bolts and wonder how the lash-up would handle 57 horsepower in corners. Answer: easily.

Following the roads which swing and bend through the San Gabriel Mountains far above the Los Angeles smog-blanket was no sweat whatever. The twin feels Gibralter-steady and predictable. It's very responsive, if not cat-quick; sure-footed, if not resolutely tenacious. Most of all, the R 75 is solid. Unlike some touring machines, one doesn't herd the machine through corners. Mind and eyes can concentrate on the road ahead without getting alarm signals through the seat of your pants. The 750 goes where you train it, and the bike gives the rider much more of the old extention-of-yourself feeling than old BMWs.

Like all BMWs, gearchanges go through with a kerplunck rather than a kersnick. Shifting may not be the fastest, yet it's positive. The R 75 changes gears more smoothly than earlier models. But, like the old BMWs, the left leg can rub against the left-side air intake tube. Ditto the right leg and tube. This is sure to happen if the rider has big legs or wears heavy boots. After a while, however, the rider adjusts to the presence of the tube. Though it still touches the leg, the rubbing isn't bothersome.

The R 75 didn't wobble a bit down the road, or shake its head. Some of last year's new BMW's had an unwanted and unsettling shake, which unnerved a few BMW faithfuls and caused much trans-oceanic head-scratching on BMW's part. Theories abounded. Some thought the culprit hid in the bolt-on rear section of the frame. Or tires. Or frame geometry. Or forks. Or rear suspension units. In truth, the shaking bug lived in the front forks. Unless the shock action in both legs were the same and the legs carefully set up to prevent wheel cock, the quivers might appear. In the 1971 models the front fork legs are very carefully assembled and the internal tolerances matched exactly, so that the forks work the way they were intended.

Why BMW designed forks with more than eight inches of travel still remains something of a mystery. The forks are replicas of the ones mounted on BMW ISDT machines; perhaps the BMW automotive-types who designed the forks figured what works well off the road makes it on the road too. And the suspension does soak up the road jolts. Up in the mountains, small rockfalls scattered the road, and the R 75 glided over the debris without pitching the rider around in the saddle. About 300 miles can be covered in a sitting, and the rider can get off the machine without aching in back, arms, legs, shoulders or eardrums. Over a long haul, the wide saddle makes a little too firm a perch. After 250 straight miles, the human sitter begins to get a bit sore.

Flex your muscles if you're going to pull the machine up on its centerstand. You'll need biceps the size of your neck to lift the machine up without strain. If there's a hundred pounds of baggage onboard, you're in for some heavy grunting. Riders of ordinary size and strength will use the sidestand. The R 75 (along with other big touring machines) ought to come equipped with a ride-off stand on which the machine can be set with little effort. The placement of the BMW sidestand could be better. It's not too difficult to ride off on the R 75 with the sidestand down, because the left cylinder conceals the arm in its extended position. Forgetting to retract the stand can lead to some pavement rash.

At night the BMW gives off a blaze of light up front, and the large tail light insures that the machine can be seen from the rear. Seventy mph on a dark superhighway won't overdrive the headlight. On a two-lane highway where the dimmer switch must be used every minute or so, switching becomes a nuisance. The left hand must be turned on the grip, and the thumb must be cocked to reach up and out toward the dimmer lever. Thumbing the switch from its topmost position (high beam) back to the low beam stop is an awkward move at first and an annoying one later.

The R 75 makes sense. On balance, it's worth the extra $300 over the R 60/5 because the R 75/5 has a broad spread of power, which the R 60 doesn't, and the 750 comes with eleven more big horses inside the engine. BMW designed the 750 for American consumers; large, heavy people who outfit their machines with bags and shields and other goodies. In these circumstances, 750 power pays off, and that is exactly what the R 75 has.

The R 75 will cover big spaces without beating or exhausting its rider. If the R 75 is something less than bikedom's most comfortable touring rig, the R 75 is also a lot more fun. Compared to pure tourers, the BMW—still the original smoothie—may be a little harder on the hide, but it's no truck. And the R 75 will cover space in a big hurry. If the machine can't fire point-to-point as fast as some superbikes, the BMW can run without adjustment and haul its load in comfort. Compared to superbikes, the BMW 750 is a little less sporting, but more fuss-free: no chain hassle. Middleground. BMW figures its a good place to be. And the R 75/5 makes a convincing case. ◉

THE PLEASURES OF (ALMOST) OWNING A BMW R50:5

BY ALAN ASPEL

Powerhouse – one of the sticking-out bits.

"TAKE IT AWAY, PLEASE", said this tall gentleman, ushering me to a bristling BMW R50/5, gleaming in the early spring sunshine. "And we don't want to see you again until there are 1500 miles on the clock." Which wasn't strictly accurate, since it had to go in for service at 600. And, as it happened, there were over 2500 miles to its credit by the time it was returned, more than a trifle reluctantly, some six weeks later. Six weeks. I ask you. We're usually congratulating ourselves if we can hang on to a machine for more than a week.

To all intents and purposes, I was the owner of the R50/5. It had but one mile to separate it from complete virginity, and, like every virgin should, it responded to the right approach and, with experience, eventually gave of its best.

The various press machines (which this was to become) have to be run in, like anything else, but unlike anything else have to become accustomed to a wide and many-varied series of riders. It was to be my pleasure to run it in, covering, as I do, a fairly high mileage – 80 a day. Chas Coombes at Slough does the fettling of the press machines, and it's always good to call on him, enjoy a cup of coffee and ride off on something expensive, especially when it's at someone else's expense.

The 1972 range was introduced at the Show in January, and didn't receive the warmest of receptions, simply because of the restyling, and for no other reason. The power unit remained almost as traditional as ever, and the /5 series which we had come to admire (after having to rethink our approach to BMW) proved among visitors to the stand to be eminently attractive – but for the tarted-up tanks and those chromium-plated side panels beneath the dual seat. Someone said that they looked like an elderly Jawa (speaking of the tanks, this is), somebody else thought they looked elderly BSA, and several shook their heads and said that Things Would Never be the Same. Which they are, of course.

The American market was blamed for the appearance of the 'bikes, and certainly the trans-Atlantic purchasers got a blasting for, it seemed, asking for smaller tanks. They didn't ask for them, but they got them anyway, although it's possible if you want (and many will) to buy the 1971 fuel tank for £6 or thereabouts – as an optional extra. Size difference is 3 gallons compared to 5½, and it's often, perhaps unkindly in view of their enthusiasm, said that our US cousins don't do long-distance riding. It looks better, said BMW, although there weren't many around at the time who agreed with them, including me. But now, after six weeks of the closest companionship, I'm ready to say, if necessary, that not only are the new BMWs the most expensive range of machines, but also the most handsome. And the fastest, too, if that came up in the argument. . . .

Those chrome side panels can be left off, should the owner wish, and they might. They may add a little brightness and depth to the machine, but that's about all. Not for me.

I'd only ever run-in one new machine before, and I'd rather not talk about that, except to say that if you had thought about buying a lightweight two-stroke of Czech manufacture – don't. My instructions with the R50/5 were to keep it below, or not above, 3500 rpm, about 50 mph in top. Oh, Gawd, the wrestling with conscience and wrist. So I tippy-toed back into London on the inside lane of M4, riding with index finger and thumb, lips pursed, aching to get out there among 'em.

Bristling with apparatus and engine – the sight well-known to many a motorist.

With 25 miles recorded, sitting in heavy traffic on the Embankment and, naturally, on the outside of the stream, it stopped. It was then that I had to face reality – no electric starter. It had been ordered with one, but someone at the factory didn't read English. Or something. It wasn't that I minded kick-starting a motorcycle; obviously, I've been brought up on a diet of kick-starting (people have come from far and wide just to catch a glimpse of my right thigh) but one had come to expect this kind of luxurious equipment. Laughing nonchalantly and pretending that I'd intended to do this anyway, I pushed the 'bike to the pavement's edge and sat there, kicking my heels and whistling nothing in particular. But with gritted teeth. A kilo has been shaved off the 1972 flywheels, giving very much improved acceleration and having no ill effect on tick-over, except when new and tight. That's all it was, although the symptoms at the time indicated fuel starvation. A kilo is 2.20 lb, so the saving was considerable. On this, the smallest of the range, the lighter flywheel plainly makes its effect more felt than on the R60 or R75. (I rode a '72 R75/5, when returning the 500 to BMW. My goodness.)

It stopped again with 32 miles gone, but this time I was half expecting it and understood its reluctance to restart. And it didn't happen after that.

Cosmetics are the most apparent alterations to this year's machines, although there are one or two minor changes around the 'bike. The side-stand now becomes spring-loaded so that one has to be careful to place it just so, to avoid the danger of a 'bike lying on its left-hand barrel. The dualseat is deeper, and infinitely comfortable. Around its rear runs a chromium-plated bar – which becomes the means of manoeuvring the machine into narrow parking spaces as well as a grab handle for the pillion. No, it isn't a cissy bar.

An inconsistency among BMWs is the quality of their brakes. One machine will halt its progress with rider and passenger sitting astride the filler-cap; another will have them both digging in their heels. That's an exaggeration, of course, but a good stopper really *is* a good stopper, and all seem sensitive to dampness. The 500 was not especially good to begin with, but with the linings bedded down after only a very few hundreds of miles, they were the equal of anything. I hope the factory doesn't decide to slap a disc on them.

No air slide or other cold-starting device was fitted and, since this was late January/early February, I was worried that an early morning might prove troublesome. Only once was this the case, and that when I wheeled the machine from the garage in a heavy downpour (didn't it rain and rain and rain . . . ?) and left it while putting on my extra, wet-weather gear. The HT leads were dripping wet, but when wiped dry, this did the trick. Or maybe it was the swearing. Thereafter, it would always start first kick, provided the carbs were generously flooded. One point to remember, for those with a lack of height, was to find a kerb wherever possible and, when pulling in to a garage forecourt, to make a point of stopping to the *right* of the pumps. This applies only to those without electric starting or legs.

Perhaps the most surprising aspect of BMW motorcycles, at least to the uninitiated, is the manner in which they take themselves around corners. Quite unlike anything else and, should you hear someone claiming to have worn out the base of the rockerbox covers, shake him by the hand, for he is a man of courage – and not a little imagination. And, while you're at it, ask to see his pointed feet. The rear dampers have been stiffened up. I had them on the softest setting of the three, and it still gave a hard ride. Front forks were a lovely compromise and you'll always be the quickest through a bend, if you go about it the right way. For sheer throwaboutability, BMW are unbeatable. Which brings me to the tyres.

Drifting tyres

AN AWFUL LOT OF PEOPLE tend to overlook, or underestimate, the importance of tyres. Many's the machine I've stepped off (not literally) with some rude comment about the tyres. Especially the early Japanese editions which would eventually have solved the population explosion. Metzelers, as fitted to BMs, exactly fit the character of the machines and, with a WM3 rim at the rear, will defy most liberty-takers. Previous models had WM2, you'll remember, and (as the actress was heard to remark to the ecclesiastical gentleman) that extra half-inch or so makes all the difference.

There's almost as much fable surrounding BMW as Rolls Royce. A certain, very famous,

THE PLEASURES OF (ALMOST) OWNING A BMW R50:5

television personality known and seen to be a motorcyclist declined the offer of a ride on a BMW since, he said, he didn't like "those bits sticking out the sides". What bits sticking out the sides? I asked my BM-owning, all-wise friend. "He means the footrests", he said. I thought he was referring to those big, finned, hand-warmers.

I never thought I'd live to see the day, but am able to say (and this was confirmed by my BM-owning friend), that the tyres distinctly drifted in the wet. Not enough to make you wish you'd gone by train, but just enough to be satisfying. Not to be encouraged, either, unless you're a masochist.

"Perhaps the most surprising aspect of BMW motorcycles, at least to the uninitiated, is the manner in which they take themselves around corners."

Within a week of collecting the 'bike from Charlie Coombes, it had the requisite 600 miles on the clock. I got it to the shop with 599.98 registering, sorry that I'd lost the use of it for a day and even sorrier that it wasn't sparkling as it should have been. But then we'd had about seven inches of rain that morning.

Initial servicing comprises an oil change, tappet adjustment and a general look-over. It needed only the minimum of attention, since nothing had gone amiss or astray. It had loosened up a lot, and was becoming much more of a pleasure to ride. The 3500 rpm limit still applied, but was getting easier all the time; the brakes were at their maximum efficiency; and the tyres well scrubbed. The time when I'd be able to take it to 6600 approached and I looked forward immensely to really putting around as it should.

The last time we rode BMs, I faulted the lights, but couldn't say the same about these. A great, white beam shimmers from the 160mm headlamp with its core surrounded by a soft spread of illumination, sufficient for any pace of night-time travelling. Electrics were always a strong point on BMW. Bosch, as usual, provide the equipment: 12 volt, crankshaft-mounted alternator, 15 amp hour battery, to cope with electric starting (but not on my model), indicators, lighting, ignition, stop light. The latter is operated by both front and rear levers, which is a commonsense arrangement, although nothing new, despite the recent claims of a Japanese concessionaire to the contrary. Other makes, I know, have featured it in the past, my 1955 Vincent among them.

Fuel consumption, during running-in, was 67 mpg, which I

Surely one of the most elegant profiles in the business?

thought to be a respectable figure. Even when functioning fully, it returned a consistent 62 and this, so I'm told, is the usual tally regardless of how they are treated.

Resulting from a visit to the Berlin factory was the knowledge that it *is* possible to make gear-changes without the dreaded klunk. It's simply a matter of bringing the lever up against the stop before using the clutch lever – engine-speed clutch, by the way. It always demanded concentration and not even at the end of my several weeks was I able to make a silent gearchange without a conscious effort. But possible, it definitely is, and very satisfying when achieved. Lovely little gearbox, too – very light to the touch.

I got the GO sign from Charlie Coombes, and took it for a thrash. First question among the younger set is invariably "Wotllitdo?" It's not the fastest accelerator, nor has it the highest terminal speed. But it has nothing to be ashamed of on those counts, and has a unique quality which lesser machinery may well envy – the ability to cruise, effortlessly, quietly, without strain on either man or machine, with the maximum efficiency. It's never any trouble, that's what counts – unless you consider a Sunday morning shampoo and the occasional tappet adjustment a tiresome chore.

I got 90 mph, exactly, out of it – speedometer reading. Inspector Alex Smith of the Metropolitan Police, six feet four inches and 17 stone in his size 12s, could manage only 75. Poor little beast.

Oh, the ecstasies of BMW owning! Garage attendants, you will find, will say, "How much?". "Almost £1000", you reply. "Or so I believe." "Blimey", they'll reply. "That's a bleet'n lot of money for a motorbike." "Is it?", you'll reply. "Mmmm (yawn), I suppose it is."

ENGINE: Twin cylinder, air-cooled four-stroke. Capacity, 498cc. Bore and stroke, 67mm × 70.6mm. Compression ratio, 8.6:1. Electric starter optional extra; one-piece crankshaft; Eaton oil pump. Single, dry-plate clutch with diaphragm spring.
FUEL: Tank, 3 gallons, including one gallon reserve; sheet steel.
ELECTRICS: Battery 12v 15 a.h. Crankshaft mounted alternator. Power 180/300 watts. Coil ignition, automatic advance. 160mm diameter headlamp, built-in parking light.
TRANSMISSION: Four-speed gearbox, left side foot pedal, up-for-up. Gear ratios: 1st, 2.835:1; 2nd, 1.875:1; 3rd, 1.364:1; 4th, 1.091:1. Rear drive, shaft.
FRAME: Welded steel tube, double loop; rear section bolted on.
SUSPENSION: Front, telescopic fork with 214mm stroke. Rear, hydraulic damped with 125mm stroke, three-way adjustable for load.

WHEELS: Front, 3.25 × 19; rear, 4.00 × 18.
BRAKES: Both drum. Front, 7 in, tls. Rear, 7 in, sls.
WHEELBASE: 54.6 in. KERB WEIGHT: 408 lb. PERFORMANCE: 90 mph maximum; 97 mph claimed, rider prone.
ACCELERATION: 0–60 mph in 9.8 sec (claimed).
PRICE: £949.00, including tax.

Touring

BY FRANK CONNER

The Superstition Mountains and an R-75 BMW

From Prescott to Douglas,
All the ranges I know . . . I drift with the wind;
No one cares where I go.
—Old Arizona Trail hand song.

The Swedish go-go-dancer and I were the only customers at Roy's on that foggy Monday morning in March. From Roy's you can usually look across the Pacific Coast Highway and watch the surfers taking off at State Beach, but this morning a heavy fog was shrouding Santa Monica Canyon, and only the dim outlines of the breakers were visible. The girl had lazed away a couple of hours over eggs and coffee, talking about her dancing, and she said, "It's a bad place where I work, but I get along okay with the customers and there's never any trouble—and the owner is a good boss, which makes a big difference." Then she said, "When are you going to split?"

I glanced at the clock, slugged down my fourth cup of coffee, reached for some money, and said, "Right now."

From behind the counter, Mac asked, "Where you riding off to this time, Frank?" Mac is a short, stocky, ugly, redfaced, gravelly-voiced individual who other people instinctively like: he can restore order to an impatient jam of rambunctious high-school surfers effortlessly, with a smile.

I answered, "Don't know for sure, Mac, but I think maybe I'll go to Silver City, New Mexico; never been there before, and I've been wondering about Route 60 and Route 70 for a long time. That ought to be some good riding."

Outside, the 1972 R-75 BMW awaited, its black paint gleaming wetly in the heavy fog. This year the gas tank has shrunk down to 4½ gallons, completely changing the looks of the motorcycle, making it seem longer and lower and far more graceful. But now they've added a bunch of chrome. The new gas tank has chromed sides, with indentations painted black, and there are

PHOTOGRAPHY: BILL DELANEY

chromed side panels, with the same kind of indentations. I could look at the bike and almost hear the marketing guys talking: "But we gotta make it more exciting visually!" And somebody else says, "Well, you know how those American riders are—chrome, that's what they like." On some bikes you can use that approach, but not the BMW. It was so clearly designed for stark, functional simplicity from the wheels up that the chrome really looks out of place.

The BMW saddlebags were loaded for a casual week's ride, which is to say that I wasn't lugging along much stuff. There was the Parr rainsuit (which is thin and tough and fits over leathers and keeps the water out, but doesn't blow up like a balloon in the slipstream); a sweater; thin, two-piece longjohns that are *warm*; skiing gloves; spare underwear and socks; manuals (owner's and shop); spare bulbs, cables, and gaskets; shaving gear; and a road atlas.

I can get away with such a thin load because I never fool around with shirts and pants on a trip—instead I wear a leather jacket and pants. While riding in the desert, enough air circulates through the leathers to keep me comfortable, as long as I keep moving. In the mountains I add the longjohns and sweater. Each night I clean off the day's accumulation of bugs and dirt with a wet washcloth. I'm convinced that leathers save skin and bones if you get off a bike at high speed. And wearing leathers sure cuts down on the amount of clothes you have to take along. I never felt very sociable in black leathers, so now I wear some wild tie-dye stuff that you can't get anymore. But there are plenty of interesting high-color smooth and roughout leathers available to the touring rider who wants to look different.

My lightly-loaded BMW was all set to go, so I rocked it off the centerstand (the only easy-to-use standard centerstand that I know of), and climbed aboard. I am 5′8″ tall; the saddle is 32″ off the ground with me on the bike; so I didn't have any trouble planting both feet on the pavement. Just then a passerby walked up, studied the machine, and said, "Boy, that bike must weigh a ton. I'd be scared to ride it."

I smiled cockily and said, "It's no problem." I didn't tell the man that the Bayerische Motoren Werke had gone to great trouble—like using forged-aluminum foot pedals, and making their aluminum castings with lots of webbing for strength without weight—to make the BMW massive-looking and strong, but light. Loaded, the machine weighs 450 pounds, which is not bad for a 750cc touring bike.

I fed in a little choke and touched the starter button. The starter motor rotated everything quietly and efficiently (there is also a kickstarter lever, so that all is not lost if your battery breaks way out in the middle of nowhere). The engine caught, and it idled quietly and efficiently. Starting a BMW is anticlimactic: nothing much seems to happen.

I pulled the clutch and punched the shift lever down into low with my left foot. The gearbox said "click"—quite a change from the old BMW "CLUNK!" I let out the clutch and headed toward the streets of downtown Santa Monica.

The gearing is excellent for aroundtown riding. Low gear will get you to 40 mph, so you can ride either in low or second. The engine comes on at about 3500 rpm and redlines at 7000.

Low-speed handling is light and smooth. In fact, this particular R-75 leaned into the turns more smoothly than anything else I have ever ridden. Some bikes lean over in a series of small uncoordinated flops: the front wheel falls over; the rear wheel catches up; the front wheel falls over some more; the rear wheel catches up . . . But not this BMW. It would lean over with one steady, smooth, precise motion that was beautiful to feel. I swooped around corners in Santa Monica for awhile, just enjoying the feel of the motorcycle leaned over.

When I was approaching traffic lights at low speeds, I found that I could lock up the double-leading shoe front brake fairly easily, but at higher speeds I didn't have any trouble. Too, the DLS front brake wouldn't prevent the bike from rolling backward if I were stopped facing uphill; then I had to use the leading-shoe/trailing-shoe rear brake.

I caught a freeway heading east. The coastal fog soon turned into bad smog. The air was heavy, and it tasted sour; soon my eyes and throat began to burn.

The BMW had found itself a home on the freeway. In fourth gear, the engine idled at 4100 rpm (70 mph). In third gear, 70 mph was about 1200 rpm faster. There was almost no vibration anywhere, and I could see everything in the mirror quite clearly.

As I rolled along, a young guy in a Porsche got interested in the motorcycle. He pulled up alongside, checked the machine carefully, dropped back and studied it from the rear, then pulled up to look at the profile view again. Deciding to be friendly, I said, "Hellooo." He stared at me deadpan, punched his accelerator, and disappeared into the traffic ahead.

The seating position on the bike was pretty good. I had to sit on the front part of the saddle to reach the handlebars, but I wasn't jammed up against the tank, and the front part of the saddle was very comfortable. Everything was great right down to my shins, where everything *stopped* being great. There is a 32mm Bing carburetor at the rear of each cylinder. This year the carburetor has a pressure-equalizing chamber up top, which works like a champion but lives right where your shin ought to go. If you have to stop quick, the chamber puts dents in your shins (even through boots). The shift lever and brake pedal got in the way of my feet. I had to ride with the balls of my feet on the pegs, or with my feet hanging off the ends of the pegs. As is, that part of the motorcycle just does not fit me.

Whoops! I had reached the first section of bad pavement grooves. The BMW danced on the grooves—not frighteningly, but uncomfortably. I screwed down the steering damper to see what would happen. It solved the problem; the bike started tracking straight. Emerging onto plain old concrete again, I unwound the damper.

At Banning the smog went away as if it had never been invented, and I was riding in bright sunshine. The air tasted fresh. I could see the mountains. Boy, could I see the mountains! The Southern California mountains and deserts were put there just for the pleasure of motorcycle riders; you cannot get enveloped by the sweeping grandeur of those large-scale landscapes if you're riding in a car—not even a convertible. For a long time I just rode, feeling the power of the countryside all around me.

But something had been puzzling me. The BMW has this tremendous long-travel suspension; 8½″ of travel up front, and almost 5″ in back, which should be enough to handle almost anything. Yet at 70 mph you could feel all the small bumps; the suspension wasn't reacting to them. The machine had about 1600 miles on it when I got it, so the suspension should have been broken in. On a hunch I ran the bike up to 95 mph, and the small bumps disappeared. Could those BMW designers possibly have designed the machine to give the best ride at 110 mph with two people aboard? Impossible! But . . . what about the Autobahn? The whole concept struck me as being hilarious, and I rode along at 95 for awhile, laughing hard.

Many big trucks were sharing the freeway with me—doubles rigs, and weird-looking chemical tankers, and lowboys toting everything from yachts to ribbed rubber tires fifteen feet in diameter. At first I had passed the trucks without any problem, but now the desert winds were beginning to hit me, and the trucks were creating turbulence—the kind that yanks you forward, makes the bike wag its front end and wobble, and moves you sideways suddenly, two or three feet. Since I wasn't going to be doing much precision cornering on this freeway anyhow, I wound up the steering damper again, which made all but the worst turbulence disappear. When passing moving vans in really vicious crosswinds, I reverted back to the old trick of accelerating and weaving from side to side

in my lane as I approached the truck, which killed the effects of the worst turbulence.

Now I was riding in true desert. The afternoon temperature was 95 degrees. The sun was to the south of me, and the mountains to my right were bluish and indistinct, but you could still see the snow atop the knife-edge peaks. Further down the road, all but the peaks of one mountain range had been covered by drifting sand or flows of lava; for miles the horizon angled up at fifteen or twenty degrees, with mountain peaks sticking out here and there.

Past Chiriaco Summit there was more empty desert, and a stark mountain range on my left. But halfway up one of those mountains was a gigantic concrete facing, from which huge pipes ran down the mountainside to what looked like hydroelectric penstocks at the base, and there was an elaborate transformer network. But there was no sign of water anywhere around. Was it what I thought it was? Where did the water come from? Where was it going?

I was running along at about 80 mph—the BMW never missing a lick—and soon I was just floating along five feet above the road, barely perceiving the scenery and not thinking about it. On and on I went, almost in a trance. Then I saw a freeway exit labeled "Mecca". I grabbed the brakes, but it was too late; I had already passed the exit. In the midst of all that desolation, "Mecca" was the perfect name for a town, or a filling station/grocery store/saloon. Whatever buildings there are in Mecca, I hope they have old rusty tin roofs.

At the outskirts of Blythe there was an orange grove. The sight of all those deep-green leaves and orange-colored oranges hit me hard; it was just too lush. Downtown Blythe was pastel-painted and poured-concrete buildings, and for once I didn't mind them at all; they were restful to my parched eyes.

Just after crossing the Colorado River, I saw my first saguaro cactus, and then I knew I was in Arizona. But somebody may have transplanted it there just to reassure me, because I didn't see any more of them for another ten miles. I ran the bike up to 90 mph and opened my mouth. The wind bulged out my cheeks, making me feel like somebody else—or maybe like nobody else.

But later on I was running at a sedate 75 mph on a two-lane stretch of Route 60 when a car coming at me with a U-haul trailer ran off the road into the sand. The driver, looking completely unconcerned, spun the steering wheel to get back on the road, never taking his foot off the gas. He got back on the road—sideways. I grabbed the binders and downshifted and headed for the dirt on my side. But the driver, still looking bored, steered into the slide and got his rig straightened out just short of me. I stayed on the road and sped up again, but my hands and feet and knees quivered for the next five miles down the highway.

Just outside Salome, a Stearman biplane sitting on the most haphazard-looking taxi strip I have ever seen, was drinking crop-dusting chemicals from a tanker. Paralleling me about twenty yards away in the desert sand, a high-school student was riding his trail bike, sitting loose in the saddle, bouncing up and down. I beeped my horn. His head flicked around; still bouncing along, he gave me a big wave and a smile.

I slowed for Salome, a small, ugly town with a lot of character. Accelerating out the other side, I was closing fast on a big semi ahead of me. There was a car coming in the other direction; it was going fast, and it was already very close. But I whipped the BM into the left lane and tucked in on the tank and let it hammer.

The semi was crowding the centerline; he probably hadn't even seen me yet. The car had seen me, but he was crowding the centerline too. His front bumper was getting big. I had to make it, and it was going to be very close. The bike flew along, rock-steady. When I guessed that my rear fender was even with the truck's front bumper, I threw the bike to the right, sliding past the car's left front fender. I smiled, realizing that this BMW was truly a motorcycle.

Later I felt guilty about maybe disturbing the truckdriver by showing up on his front fender like that (no regrets about the car), until I got another ten miles down the road. I topped a rise, and right in front of me were two semis heading my way, side by side. There was just no room for me at all. So I grabbed the brakes and came to a full stop on the side of the road. The truck in my lane edged over just enough to miss me; mighty obliging of him.

I raced with the sun to see if I could get to Wickenberg in daylight. The mountains up ahead were growing softer and hazier, and the air was getting cold. In my mirror I watched the sunset; a red and purple and orange sky. I pulled into Wickenberg a few minutes later, but by then it was pitch black dark. I went to bed easy in my mind about the motorcycle and the trip.

The next morning I rode alongside the Hassayampa River as it meandered sloppily along a wide sandy bed cut through fragile hills, with shade trees growing on the banks—a magical place. Then I took the freeway through smog-tinted Phoenix—along with a lot of going-to-work traffic—and picked up Route 60 on the other side. Toward Apache Junction I could see the Superstition Mountains up ahead. From that direction they bulge strangely, and the peaks stick out at odd angles. Stopping for gas at a filling station, I asked the attendant, "Do people still shoot each other back in those mountains?"

He looked at me strangely. "I wouldn't have believed it—only moved here a few months ago—but I guess they do. Crazy Jake came by the other day, and he said somebody took a shot at him back in there a couple of weeks ago."

I paid him, fired up the bike, and rode it down the highway until I had a full side view of the Superstitions. Then I parked the bike off the road and sat down on a rock and stared at those ominous mountains baking in the sun. They got their name because the Apaches feared them. But maybe the white man has even better cause for fear. The mountains brooded, and their influence weighed heavily on me; I felt uneasy. Legend has it that one of the Peralta brothers (big landholders in Mexico a long time ago) ran short of operating money and brought a crew north to look for gold. They found nine veins of it—wire gold in quartz—there in the Superstitions. Being greedy, they came back for more, but then the Apaches killed most of the party and sealed off eight of the mines. A Dutchman, Jacob Waltzer, supposedly found the ninth mine near Weaver's Needle, a landmark that looks something like a high-crown sombrero. If so, the secret died with him. Other people have been hunting the Lost Dutchman mine ever since. The area isn't very big—maybe forty miles across—but it is a waterless maze of mountains and canyons, and the temperature in some of those canyons reaches 140 degrees in August. Many people have gotten lost and died of thirst in the Superstitions.

Others have been shot from ambush by prospectors gone crazy. I first went looking for the Dutchman mine fifteen years ago, taking two hardcase friends with me, and we were all well-armed. Three people had been murdered in there just before we went in. Now as I sat staring at them, those mountains looked and felt as evil to me as they had fifteen years before.

The road followed the Superstitions to Florence Junction; then Route 60 began an end run around them on its way to Globe. Just past the forks I saw a paved road leading back into the mountains; the sign said, "Superstition Valley Golf Course." I guess that's the beginning of the end. Ten years from now people will be crowding into luxury motels all over the Superstitions, and there will be fake gunfights every night to amuse the tourists. Perhaps it is just as well; those old mountains have soaked up enough blood.

Running through desert, I topped a rise, and there was Superior—a mining town—spread out in the valley below me. The road got curvy on its way down, and I overtook a stack of fifty cars and trucks and campers crawling along. I took to the centerline and left that collection of vehicles behind me, feeling like a free man.

On the other side of Superior the road looked as if it would run smack into a sheer cliff, but at the last moment it angled to the right and began climbing hard. I followed it. Up the crumbly mountains, weatherbeaten and cracked, I rode past vividly-colored strata of rock—white, pink and orange. Queen Creek Canyon, a narrow, dry gorge, ran right beside the road.

Later I rode by Bloody Tanks Wash (and I don't think I want to know about that place), toward Miami-Globe. The first thing you see in Miami is a huge pile of mine tailings and a giant smelter stack belching chemicals and smoke.

At a restaurant in Globe, the waitress said, "Been real dry in these parts the last two years. So dry that a lot of cattle on the San Carlos reservation have died of thirst."

I told her about the Seven Years' Drought in West Texas, when newcomers to El Paso had built houses in the mouths of arroyos and lived there for maybe five years before the first rains came and washed them all away.

Just beyond Globe the highway forked. Route 70, the road I was supposed to take to Silver City, angled off to the right. Route 60 went left and headed north into the San Carlos reservation. For some reason I turned left. The countryside was dirt and low shrubs—maybe a little greener than pure desert. Then the road climbed up to a pass, and suddenly there were pine trees and other evergreens. They looked a little scraggly, but they were the first pines I had seen on this trip.

Running along at 70 mph, I started catching hard, medium-size bugs. Zing, splat; zing, splat. They stung my chin, and gradually they covered my face shield. One bug splattered right on the edge of my shield and ricocheted up into my open mouth; he tasted bitter.

Now I was at 5000 feet elevation, and the air was cold. I passed an Arizona Highway Department outpost, and they had a whole string of longjohns hanging on the clothesline. I stopped and went behind a tree and pulled on my own longjohns.

Just past the crest of another hill, there was a steep cliff and a tremendous valley spread out in front of me. I stopped; the sign said, "Salt River Canyon." I'd never even heard of it before, but I was enchanted. At the rim of the canyon I sat on a weathered piece of lava with Spanish bayonets growing on either side of me and some prickly pear behind. Down below, the olive-green Salt River was shooting rapids and making white water about a mile away, and the air was filled with the dull roar of the rushing water.

I climbed aboard the BMW and went a hundred yards and found that the canyon I had been studying was only one of a huge network. Though not quite as large as the Grand Canyon, they were fully as spectacular. Through the ages, the Salt River seems to have chopped off the sides of great mountains, leaving caves and potholes in their faces. The river has cut its way down through all different kinds of rock; you can study those layers either from a distance or from five feet away, because the road switchbacks right down to the bottom of the canyon, crosses the river, and winds its way up the other side.

Some of the rock layers are softer than others, so the river has cut ledges and steps; veins of hard rock stick out vertically here and there. There were turnouts all along that road, and I stopped at each one to study the magnificent scenery.

I followed Route 60 up through the Fort Apache reservation and into Show Low, where I headed west on Arizona 77 into Snowflake. Then, right out in the middle of nowhere, with no houses or other buildings or towns or *anything*, I rode past a huge papermill running full tilt. Why was it there? Another mystery.

The road began to climb mountains, and then I was in the Tonto National Forest. The woods were thick, and the pine trees were tall. The afternoon sun flicked on and off between those trees so rapidly that it was like staring directly into somebody's strobelight on fast flash. After awhile it makes you dumb, and you can't think anymore. There were groups of birch trees scattered among the pines, and many small ponds, and little patches of unmelted snow and ice in the shadows.

To the north was the Mogollon Rim, a huge mesa that drops off abruptly into the Tonto Basin. Zane Grey had a cabin in the Basin, and he wrote about the range wars that took place there in the 1800s.

At Payson I turned north on Arizona 87, into more forests. Then there were still, cold lakes surrounded by meadows, and to my right were weathered boulders stained chartreuse by lichens. To me that whole scene looked just like the tundra country of Northern Canada.

Then I hit twenty miles of the worst paved road I've ever ridden. It was colored deep red, and it had potholes a foot deep, and bad (horrible) washboard that you could not see in the late afternoon. The bike was bottoming its suspension, waggling its forks and bouncing all around. I couldn't blame the BMW; I think that any street motorcycle would have reacted the same way. When I could see them, I rode the worst sections standing on the pegs.

Then the setting sun lay right above the road. But this sun was not red and dim like a setting sun ought to be; it was bright yellow, and running full blast. I had no visor on my Magnum, and the sun blinded me completely; I had absolutely no idea which way the road went next. In a panic I slowed to ten mph and tried to figure out what to do. Finally I discovered that by covering my shield with my left hand and peeking out between two fingers, I could see a little bit. I rode like that for fifteen minutes before the sun went down. It was dark when I got to Flagstaff.

I checked into a motel, found a restaurant nearby, and ate dinner. A man walked into the place, saw me, did a doubletake, and asked, "Pardon me, but did you ride a motorcycle through the town of Show Low this afternoon?"

"Yep."

"By George, I didn't *think* there were two set of clothes in Arizona looking like that. I work for a paper company, and I was inside a store selling a man some paper goods when you rode by."

Aha! Maybe this man could solve one of the mysteries for me. "You know that papermill way out in the middle of nowhere between Snowflake and Herber?" He nodded. "Good. What's it doing there?"

He laughed. "That mill has been there since 1960, and I've asked everybody I know, including some of the management, and nobody know why it's there."

Arriving back at my motel, I saw an immaculate 500cc Suzuki twin parked outside the office. The bike had a handlebar fairing, and camping gear laced neatly to a sissy bar. While I was checking out the Suzy, its owner came out of the office to have a look at the BMW. He was a young rider named Mario De Beneditis, who worked for the telephone company in Denver. Mario was returning from a week's vacation to Los Angeles. He and I decided to hunt down a six-pack and swap lies about motorcycles for awhile.

"Well," said Mario, "I like good handling, good braking, and fast acceleration, in that order. If a motorcycle won't handle right, why buy it? And if it won't stop quick, you're in trouble. Anyhow, that's what I wanted, and I don't think you can find that combination in any other 500cc twin for less than a thousand dollars, so I bought the Suzuki. It's been reliable—hasn't given me any trouble at all. The only problem with it is that it vibrates way too much; it's got a high-frequency vibration that really bugs you."

I asked Mario how well his camping-gear rig worked on the road.

"I bought that sissy bar for the bike, and I tie my tent and sleeping bag and travel pack to it. Then those things are out of my way, and they don't get loose, and the bike isn't cluttered, and it's comfortable, but it still doesn't work very well. The sissy bar with my gear tied on acts just like a sail and catches the wind. It blows the bike all over the road, even though I keep the steering damper clamped down. You know how bad those winds get in the desert." I nodded; I knew.

"Well, one time this afternoon the wind blew me right off the road. I was going pretty fast, and the shoulder was soft dirt, but I was able to keep the bike up and get it back on the road . . ."

We swapped bikes and rode them around Flagstaff for awhile, then came back and talked until 1:30 in the morning. The next day we met for breakfast and continued to talk about our trips over coffee. I rode away from Flagstaff with a glow. When you're alone on the road, one of the nicest things that can happen is to meet another rider passing through—a good guy—and spend a few hours talking.

I left Flagstaff on Route 89A, mostly because I had never been on that road before. A few miles down the road a sign said, "Lookout Point Ahead". I couldn't see anything to look out at; the countryside was just rolling hills. Then the road dived down into a huge gorge—Oak Creek Canyon. The air had turned bitter cold, and there was a thin scum of ice on the asphalt as I rode beside Oak Creek. Then the sun hit the towering cliffs on my right, with that extra intensity of a backlighted Kodachrome transparency. The top half of the cliffs looked like granite, which had yielded only grudgingly to the water, showing gouge marks here and there. The bottom half was brick-red sandstone; it had given in gracefully and was molded and curved.

The canyon widened out, and then there were individual mountains of sandstone eroded into fantastic shapes. In the mouth of the canyon was the town of Sedona.

I headed west across a wide valley, and the vegetation died away; a bare mountain range loomed ahead. The road dipped down and crossed the Verde River into Cottonwood, with its stand of huge, stately cottonwood trees growing along the flanks of the river, as if planted there in groves.

Three-quarters of the way up Mingus Mountain hangs Jerome, clinging by its fingernails to the steep mountainside. Jerome is a ghost town: in 1931 it had a population of 20,000; now 300 people live there. It was a copper-mining town which experienced almost every kind of disaster imaginable. For example, the main ore body (which has a heavy sulfur content) caught fire, and they couldn't put it out, so they stopped tunnelling for the ore and instead dug a huge open pit down from above so they could strip out the burning ore; that went on for several years. But now the rich ore is gone, and so are all of the miners.

On the other side of Prescott, I ran into about twenty miles of crooked, winding mountain roads, with no scenery worth looking at. So I gave the bike a workout—it was almost never upright. I shifted back and forth between second and third gear on that 30 mph road and had the time of my life slinging the BMW into the corners. In the open curves that I could see around, I was using every inch of asphalt.

The mountains died off into a grassy valley, and then I climbed some more hills to the town of Yarnell, at an elevation of 4700 feet. At the other end of town, the land drops off in a vertical cliff; half a mile straight down is flat desert stretching westward as far as you can see. I followed a Transcon semi down the narrow winding road. He was using every inch of the asphalt—just as I had done in the mountains outside Prescott—so I just trucked along behind him. Then came a short straightaway, and I was gone. On the desert floor, the air was like a blast furnace.

There were dust devils near Salome. They were about six feet in diameter, and some of them were a hundred feet tall—miniature whirlwinds greedily sucking up dust from the ground and storing it in their fitfully-swaying columns.

Back in California, I left the freeway at the Mecca/Twenty Nine Palms exit, and turned my back on Mecca (regretfully) to ride through the Joshua Tree National Monument. The Joshua trees—yucca plants that graduated—were there in great numbers, along with every imaginable variety of cactus. The harsh mountains still looked harsh in the dim light as the Southwest staged another spectacular sunset.

That night I rode the streets of Palm Springs, watching the people saunter along the sidewalks illuminated by the floodlights tied to the palm trees lining the streets. The people wore resort-town finery, and they strolled slowly, displaying their own finery and checking out the other people's. Should someone else out-finery them, they can always stroll into one of the exclusive shops and buy better armor.

I headed for Los Angeles. Turning it on, I could only get 100 mph on the clock, sitting up. But by now the machine had 3600 hard miles on it (of which I had contributed about 2000), and it was evidently beginning to lose its edge. But I had to turn it on to learn that.

The R-75 was getting about 46 miles per gallon at 75 mph, which meant about 200 miles on a tank of gas. The machine still wasn't shaking, and it hadn't dribbled off any nuts or bolts along the way. It had run quietly and tirelessly. It was very comfortable to ride from the shins up. It was one of those rare combinations—a good bike for putting around town or for tooling across the United States. A sloppy rider can feel at home on it, and a skilled rider can get a lot of performance out of it.

The present R-75 is a far cry from the R-69 model of a few years back. The R-69 had been designed for a sidecar, and you could tell. There was a great flywheel-effect lurch when you shifted gears; to go faster, you had to wind up the twistgrip like an alarm clock, and even then nothing startling happened; and in the corners the machine felt as if it were itching to dump you off. The R-69 was a two-wheel means of conveyance, rather than an exciting motorcycle. But—as with the Harley 74—there was a contingent of BMW fanatics who loved the R-69 just as it was. It took a lot of courage for BMW to leap out of their comfortable niche and build an entirely different type of motorcycle. Happily, the R-75 justifies the risk, and it should appeal to a wide range of riders.

I pulled into Santa Monica Canyon—still foggy—and parked outside Roy's, all set to tell Mac and the Swedish go-go dancer about my trip. But the dancer was working that day, and Mac was resting up at home after having a couple of teeth pulled. Foggy or not, the waves sounded good. So I went back to the motel to grab my wetsuit and my fins. ◉

SUPER TWIN

FOUR NON-STOP HOURS ON THE TRACK TESTS THE LATEST BMW R75/5 TO THE LIMIT! JOHN ROBINSON REPORTS

AT SNETTERS!

Like Rolls-Royce and Vincent, there is an aura of myth and prejudice surrounding the BMW which cunningly camouflages the reality.

And carrying a price tag of the magnitude of the BMs, it's probably just as well.

Even so, an ordinary motorcycle it isn't. Otherwise why would complete laymen, car drivers, pump attendants and so on get curious enough about it to come over and ask questions? This is not a thing which happens on many other machines we ride, maybe it's because the untrained eye only sees engine and still more engine when it scans the R75/5's flank. Or as the car driver put it, "Why have you got half a Volkswagen in there?" To which one of course replies "It's not, it's a third of a Porsche."

Where the name BMW is concerned, motorcyclists themselves seem to fall into two categories; the out-and-out enthusiast of the marque who believes it to be God's gift and can do no wrong, and the rest who believe it's on a plane above them and that it has to be good 'cos everyone says so.

A BMW enthusiast once told me that you have to develop a special technique to get them round corners quickly. Of course this argument is unanswerable —when I replied that I'd never had any trouble, he smugly said that either I was born with BMW skills built in, or else I wasn't going fast enough!

To us, professionally, the R75/5 is just another motorcycle, and an expensive one at that. And the price, which has also always been part of the legend, is largely an artificial inflation. It's arrived at like this: the concessionaire buys the bike from the factory at a relatively low price, then he has to pay to transport it to these shores, which increases its price. As soon as it has cleared customs 18·5 per cent duty is whacked on to the price and as soon as it is sold on goes 25 per cent purchase tax (yes, you pay tax on the duty already imposed!). Then the concessionaire sells it to a dealer, at a profit of course, and he then sells it to the customer, at a 20 per cent profit. The net result is some £1200. So you can hardly say you are paying for the value of the machine when three people have made their profits and the Government has taxed it twice.

Yet the BM is well engineered and well finished, and at its more realistic German price, compares more than favourably with contemporaries like the Commando and Trident.

On the road, used for commuting and long runs, the BMW proved ideal in a number of respects. The riding position is just right, one or two up, for fast or slow riding. The seat is comfortable in all conditions and the passenger grabrail makes life a lot easier for everyone on board.

An open road really shows the 750 at its best, it's smooth and quiet, and builds up deceptively fast speeds. Top speed is not high, but the engine never has to work hard and it'll cruise at the 95 mark all day, taking hills without a murmur.

The '72 slimline styling also gives the impression of a much lighter machine and I personally like the new tank. The flashy side covers aren't quite so BMW though . . . whoops, there I go invoking the legend again!

The tank, say BMW, holds $3\frac{1}{3}$ gallons including one gallon reserve. Ours held 3.7 gallons and reserve took it slightly less than 20 miles. At an average of 52 mpg, that's under half a gallon. The slim style has been criticised as BM owners are notorious for their long runs, but it still gives a range of 190-odd miles.

The very reasonable fuel consumption is due to the new Bing CV carburettors, which seem to stabilise consumption—the mpg varies little from town riding through motorway cruising to thrashing around a circuit.

But they do have their disadvantages—they're not so good for full throttle performance and it is difficult to synchronise the butterfly valves perfectly.

Our machine suffered from

grand tourer—BMW's big one tested on road and track

The restyled R75/5 also uses Bing CV carburettors which seem to improve the fuel consumption

Layout is, as always, neat and comfor able, but instruments are difficult to re

Front stoppers are usually good on BM —ours wasn't. Note the welled r

poor starting and a rough patch of vibration at 2500 rpm, which was eventually attributed to the carburettors.

Waking the BM up in the morning was something of a problem. It just didn't want to start, and had all the symptoms of the big single syndrome. To persuade it to go, the throttle had to be held open just enough to get some tension in the cables, and a teeny bit of choke was needed. Then after about eight squirts on the starter button one cylinder would pick up and finally coax the other into life. Oddly enough, once it was going, it showed signs of being slightly weak, yet the choke was barely necessary to start it. Thank heaven for starter motors!

Two other faults showed up in the engine. One valve clearance kept opening up and made a nuisance of itself by clattering madly—a BMW mechanic tells me this is not unusual—and one pushrod tube developed an oil leak. This happens as new engines bed in and there's a special tool to tap home the seal at the crankcase end. That cured the oil leak when the engine was hot, but when it cooled down, oil started to seep out again. Obviously the cylinder and the pushrod tube expand at different rates—Triumph have always had the same problem—but why? It can't be beyond designers to eliminate a silly fault like this.

These later engines have a lighter flywheel and a reduction gear to the clutch—could this be an attempt to get rid of the clonky gearchange? BMW lovers everywhere will be pleased to hear that it doesn't, but before the letters come pouring in, let me say that a quiet change is possible. I know because I've done it a couple of times, but it takes so much effort that most of the time it's just not worth it.

While I'm grumbling about the faults, I'll include the speedometer and tachometer arrangement which isn't the most easily read instrument I've seen. And then there's that silly ignition switch, and the rear suspension is too hard.

Even on the middle setting the rear wheel would patter around corners, and the same setting proved about right for a 12 stone rider and 13 stone passenger. With five inches travel at the rear and $8\frac{1}{2}$ inches at the front, where the hell do they expect you to ride it? Across the Cairngorms?

Although there is a lot of movement at the front, this doesn't affect handling which is very good in a comfortable sort of way. In fact the roadholding matches the effortless, lazy performance of the engine very nicely, making for a pleasant, safe ride.

Through the corners you have to be fairly firm with the BM, as, given its head, it likes to run wide, but this applies to any big bike. It doesn't object to cornering on the overrun, rolling it off in a fast turn, or changing line, or any of the things one isn't supposed to do.

An important aspect of this is the tyres, nice fat Metzelers fitted on to wide rims, which seem to be just what the BMW needs. Grip was all there in the wet, too, and combined with the gentle power build-up made for a lot of confidence in the machine as a whole. And by the time the water reaches your feet, at least it's warm water!

As an illustration of its stability, I had a rare occurrence, a puncture, when the rear tyre picked up a nail. I was wiggling through heavy traffic at the time and the first I knew about it was when the BM started wiggling slightly more than I was.

As I was surrounded by heavy traffic I had to go on for another 100 yards or so before reaching the refuge of a traffic island, but the BM was almost as controllable with one flat tyre as with healthy ones! The deep, welled rims also prevented the tyre from running off the wheel.

Other points which are invaluable to the hard rider are neatly built into the BM as a matter of course. The lights are very good, the headlamp flasher even better and the horn is really loud. The tail light is large and both brakes work the stop lamp.

Under the locking seat there is a reasonably large removable tray for tools. For some reason our machine didn't have a toolkit, but I saw one down at Kuhn's BMW Centre, and it looks a cut above the average two-box, three-open-ended and a piece of bent wire for a screwdriver.

Fast, long-distance riding on the road is the BM's forte, and there are few machines to compare with it in this respect. But what seems, and is, fast on the road often looks a bit silly on the track.

When we took our performance figures at Snetterton, the BM's lap times were less than impressive. It's probably only fair to say that the rider is at something of a psychological disadvantage—one is aware that it's a very expensive machine

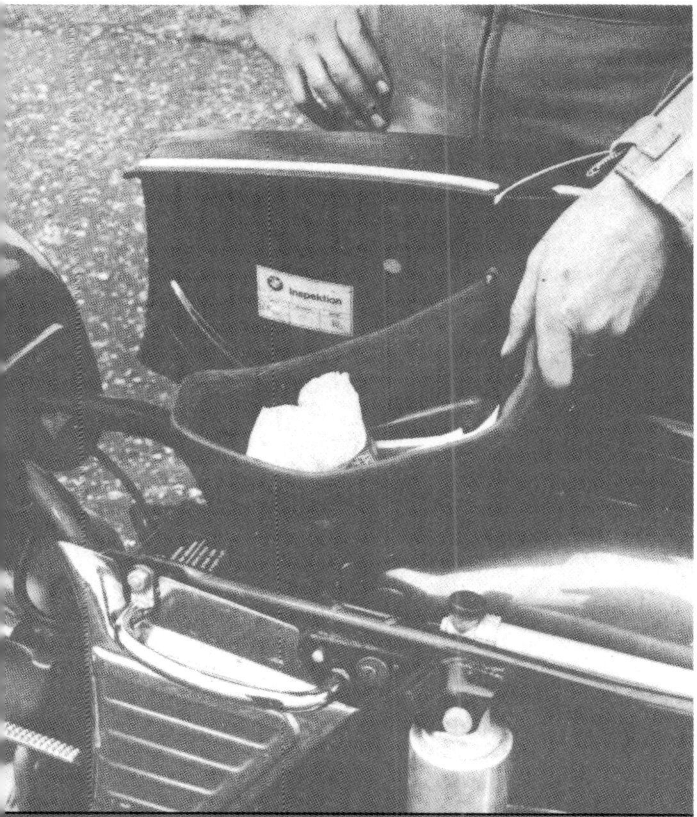

Under the locking seat there is a removable tray for tools giving access to the large capacity battery

to stuff into a bank, check the price of those front forks! And one is wary of bending it over too far in a turn, as there's the risk of an instant and permanent decoke. Actually, ground clearance is good, with only the centre stand hitting the floor.

On top of that the BM's long-legged, lazy character rubs off on the rider; it just isn't the sort of machine to be chucked about. It would be like asking Queen Victoria to a discothèque.

No, the BMW isn't a fast machine in that sense, it's not meant to be. We found its top speed to be just what BMW say it ought to be, 109 mph, which surprised us as German manufacturers are usually conservative in their claims (it's due to their version of the Trades Descriptions Act). In fact the BM may be capable of a little more, as we were running with the carbs a fraction de-synchronised and a clattering tappet

On the other hand, you can take it right down to under 1000 in top, and it'll plonk away quite happily and even pull full throttle with no fuss. The big single syndrome again.

One thing I haven't mentioned so far is the braking, and before I do, I ought to say that the front brake is identical to the one fitted to the R50/5 which I rode last year, and the R75/5 is only a few pounds heavier (11 lb. actually). Now the R50/5 had perfect brakes, one of the best roadsters ever, but the R75/5 didn't. The front brake was good, but it needed more pull and gave less feel.

This shouldn't be caused by adjustment, because the linkage is self-compensating, so maybe the linings were glazed or something.

Modifications for 1972 include a new high-pressure lubrication system and shell bearing big ends, designed to increase the life of the motor.

So while the R75/5 isn't the fastest machine around, its comfort, staying power and reliability make it ideal for most types of roadwork. Even commuting through dense traffic, while better suited for a lightweight, doesn't give the BM any headaches.

BMWs have always been one of the most elegantly styled machines around and the engineering and detail finish is as sound as you'll find anywhere. In every way this is certainly a machine with plenty of character.

The only hang-up is: does the money really mean that much to you?

ROAD TEST BMW R75/5

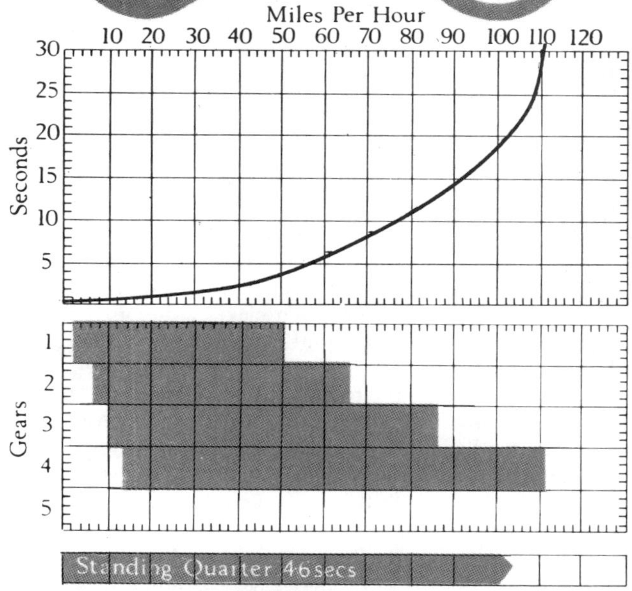

fuel consumption (average) **55 mpg**
(track test) **52 mpg**
braking from 30 mph **30.5 ft.**

SPECIFICATION

Engine: OHV, horizontally opposed twin, aircooled. One-piece crankshaft with split steel connecting rods, lubrication by Eaton pump via full flow oil filter. Carburation by twin Bing CV carburettors. Bore × stroke 82 × 70.6 mm, capacity 745 cm^3, compression ratio 9.1. Maximum claimed output, 50 bhp (DIN) at 6200 rpm, 57 bhp (SAE) at 6400 rpm, 43.3 lb. ft. at 5000 rpm.

Transmission: Four-speed crossover gearbox driven via reduction gear and single plate clutch with diaphragm spring. Final drive by shaft and bevel pinions with shock absorber in shaft. Gear ratios, 2.835, 1.875, 1.364, 1.091, primary reduction 1.375, final reduction 3.21.

Electrics: 12 volt ac/dc, 180 watt crankshaft mounted alternator, diode rectifier, 15 ah battery. Coil ignition, front/rear stoplight, flashers, and telltale lights for oil pressure, main beam, battery charge and idle (for gearchanges!) and starter motor fitted as OE.

Wheels: welled rims, front 1.85B × 19, 3.25S × 19 tyre, rear 1.85B × 18, 4.00S × 18 tyre. Brakes $7\frac{1}{8}$ inch, twin-leading-shoe at front.

Suspension: Front, telescopic forks, rear swinging arm with adjustable dampers.

Dimensions: fuel tank $3\frac{1}{3}$ gallons, overall length 82.7 inches, wheelbase 54.6 inches, seat height 33.4 inches, overall width 29.1 inches, kerb weight 419 lb.

Price: £1196 inc. P.T.

Concessionaire: BMW (GB), Chiswick High Road, W.4.

BMW R60/5

A Beautifully Engineered (If Not Exactly Beautiful) Traveling Man's Machine.

Cycle World Road Test

■ IN THE December 1970 issue of CYCLE WORLD we raved on about the new BMW R75/5, the German factory's answer to the onslaught of Superbikes from Japan, England and America. The staid BMW had taken on the character of a wolf in sheep's clothing.

Now everyone knows that a BMW is a quiet, docile *gentleman's* motorcycle: extremely dependable, practically silent in operation, and meticulously well-finished. We got a number of letters from the purists complaining about our analogy of a BMW to a Rolls Royce. We retracted our statement somewhat in the Letters column and now have to compare the BMW to a *grosser* Mercedes sedan, even in view of the fact that the BMW automobile firm makes some of the most refined and luxurious touring cars available.

In contrast to earlier BMW motorcycles, the R75/5 was starting a trend away from the stigma BMWs had carried for years. It was a very refined and cleverly disguised Superbike. No bones were made about it: BMW was out after the Superbike market.

But the R75/5 is the top of the line. The largest, fastest and most expensive model. So appealing is it to many riders that the two smaller models are often forgotten about. But after spending some time with the 500cc R50/5 and the subject of our road test, the R60/5, you won't be inclined to forget either for some time.

The words "quality" and "precision" go together even though, strictly speaking, they aren't synonyms. A precision product suggests quality, but somehow the reverse isn't always true. However, in the case of the 1972 BMW R60/5, the words might as well mean exactly the same thing. It's seldom that we see such a pleasantly integrated package as the BMW. Of course, it's not perfect, but for the purpose it was designed for, it comes as close as any motorcycle we've sampled lately.

If there was ever an ergonomically sound motorcycle, the BMW has got to be it. Everything fits the rider in such a way as to be truly comfortable. There are minor annoyances here and there according to how big or small you are, but the majority of our staffers (who range from 5 ft. 7 in. to over 6 ft. tall and vary in weight from 125 lb. to close to 195 lb.) found the R60/5 very comfortable.

It finally occurred to us after all these years why many BMW riders favor the cuff-inside-the-boot costume. It doesn't have much to do with their Teutonic bent, or with a love for things military. Rather, it is to avoid fouling the pant legs with the Bing carburetors. If you are inclined to bells or flares, beware. It's easy to catch the fuel tickler with the cuff and flood the bowls.

The view from the saddle is much like BMWs of years past. An instrument cluster in a single glass face contains the speedometer, tachometer and a myriad of idiot lights to keep the rider informed of what's happening. A single green light to the left of the combination ignition/lighting switch signals when one of the turn indicators is in operation, and yes Virginia, the speedometer and tachometer needles are still a little on the jumpy side.

Of medium rise and width, the handlebars are adorned with some of the nicest controls we've ever seen. The stiffness and crabby operation of a few on the R75/5 test machine we had just weren't there. Back to the ergonomic thing: the aluminum alloy clutch and brake levers have indentations for the rider's fingers to rest in. Each lever has a teflon bushing at its pivot point and there are spring washers between the lever and its body to keep it from jumping up and down under vibration.

A nice touch is that the levers don't have slots for cable removal (and catching dirt to wear out the cable). Rather, they have a fitting that must be pulled out before the cable and there are felt washers inside to keep dirt *out* and lubricant *in.*

In spite of the two carburetor return springs the throttle effort is only moderate, due in part to the slow throttle grip ratio which makes grabbing a second handful mandatory when accelerating to pass traffic.

A flip switch on the left hand handlebar is a three-way affair which rides in the middle position for the headlight low beam, the highest position for high beam, and is spring loaded so that it can be pushed down to flash the high beam for passing or greeting another rider. Press the switch in and the horn blows.

A similar switch on the other side controls the turn signals and pressing it activates the electric starter.

The starting system increases the weight of the motorcycle by about 50 lb. because of the weight of the automotive-type starter motor, the larger battery and the associated paraphernalia necessary. BMW engineers have found ways to help cut the motorcycle's weight, however, and the starter has a circuit

which is fed from the alternator to prevent accidental engagement of the starter when the engine is running. Even though the R60/5 was usually absurdly easy to get running by using the kick starter, the electric unit is so neat in operation it's *almost* worth 50 lb.

Being just a little on the traditional side, a couple of our staff members turned up their noses at the 1972 styling scheme. Very un-BMW chromed gas tank panels and battery covers set the bike off in a crowd, but we're not sure whether that was the way to do it. Somehow it just doesn't look like something a German would do on his own accord and we were informed that the change was made because the stylists said it

would sell more motorcycles. We wonder!

Extensive use of aluminum alloy for castings and the reduction in bulk of many steel castings have helped keep the weight down to a respectable 440 lb. w/half-tank of fuel, electric starter and all. This moderate (for BMW) weight and excellent steering geometry make the machine easy to handle around town and very stable at speed on the road.

You look at the engine and say to yourself that it's mighty tall, in spite of the low cylinders. That's true, but the bulk of the top half of the main engine casting is used as a compartment to house the starter motor and the dry treated-paper air filter element which is hidden at the end of a baffled chamber to reduce the intake noise under hard acceleration and fast running.

Additional weight reduction has been realized by the use of aluminum alloy cylinder barrels instead of the older cast iron ones. These barrels feature a perlitic cast-in steel liner which gives superior heat transfer to the cooling fins and excellent wear characteristics. Because of uniform expansion rates, very close piston/cylinder wall clearances can be used, which result in more power and quieter running.

Steel connecting rods with three-layer plain bearings at the bottom and a plain bush at the top transfer the movement of the pistons to the crankshaft. In order to shorten the engine somewhat there could be a complicated system of divided connecting rods, but BMW engineers preferred to do things symetrically downstairs and have a forged steel crankshaft with two separate throws.

Looking down on the engine from the top is a little strange at first: the left hand cylinder is slightly more forward than the right! At the extreme outside, the cylinder heads have

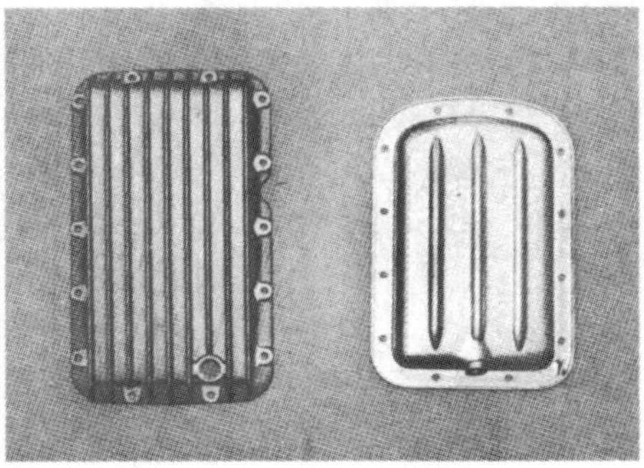

The cast aluminum sump cover on left has more bolts holding it on and is more rigid than the older, pressed steel unit on the right.

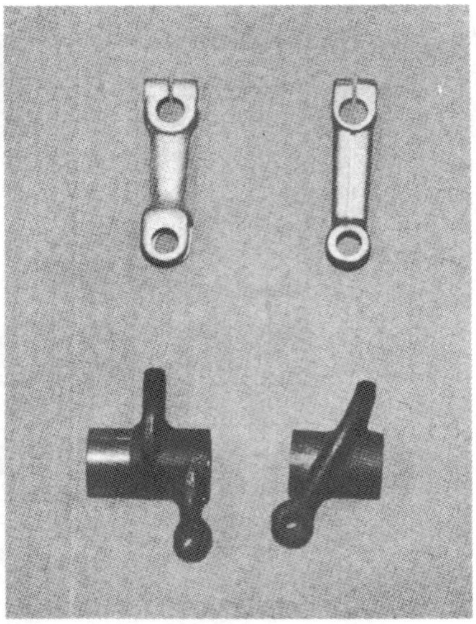

Old style brake pivot arm and rocker arms (left top and bottom) are larger in mass and weight than the latest ones.

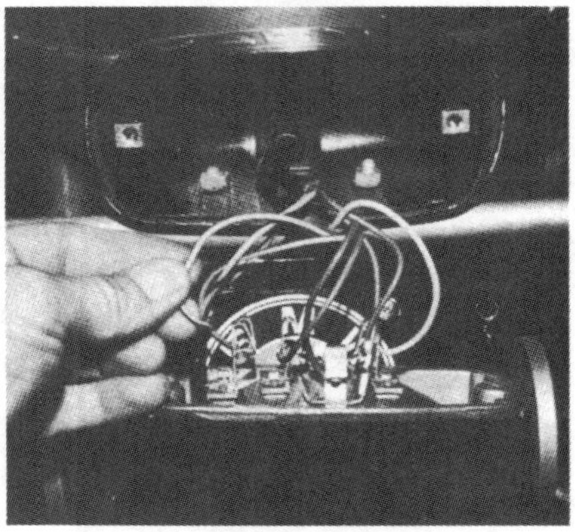

Sanitary wiring layout features spring-type connectors which eliminate solder joints.

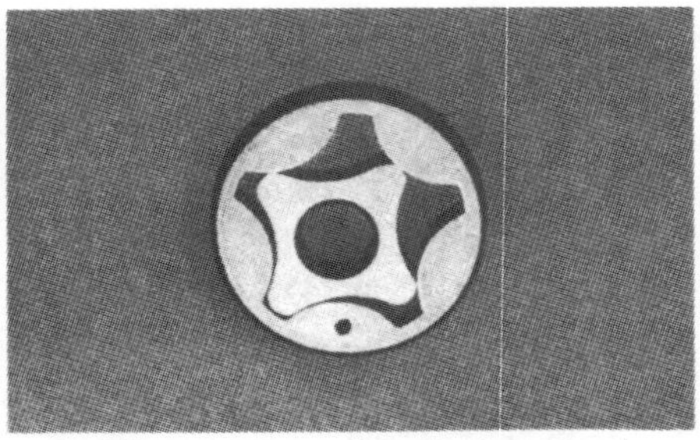

The Eaton system oil pump featuring hypotrochoid teeth.

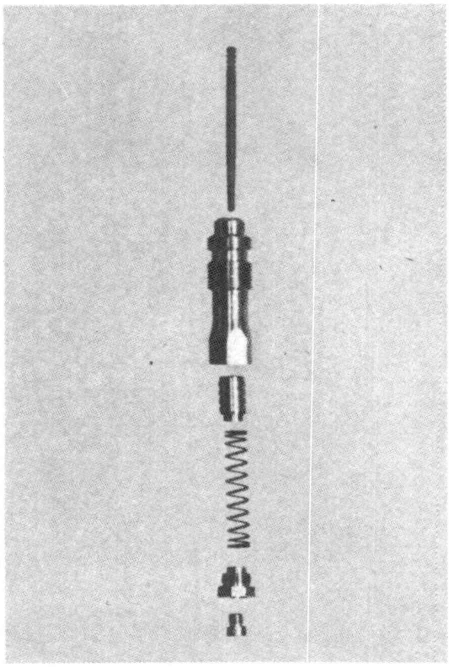

The carburetor's jet needle and accelerator pump assembly is simple and works without fault.

benefited from a redesigned combustion chamber with a shallower valve angle (63 degrees instead of 80). With this design a piston with a smaller dome can be used to get the required compression ratio and still get good breathing characteristics. Tiny steel rods go between the cylinder head fins to help deaden any valve noise.

Riding a BMW is one of the most delightfully *quiet* things you'll ever do. There is hardly any noise from the closely set-up engine and the rakishly upswept mufflers emit only a murmur, even under full throttle acceleration. A balance tube between the exhaust header pipes aids power output somewhat and helps reduce exhaust noise. Ah ha, you thought the mufflers were upswept for styling and ground clearance. They are, but because they are slanted back down toward the exhaust header pipes, moisture which begins condensing inside almost immediately after stopping the engine can drain back into the headers and evaporate as steam instead of settle and rust out the mufflers.

Some of the engine's quietness and smoothness comes from the duplex-type chain which drives the camshaft and features a tensioner blade which helps damp out vibration too. If you remember the older BMW with the camshaft *above* the crankshaft and driven by a gear you might think that the engine is higher. It is, but there are a couple of reasons: the higher cylinders raise the ground clearance for improved cornering speeds made possible by today's better tires, and the camshaft receives better lubrication being down near the oil supply.

Oil which lubricates, cleans and cools the engine is circulated by a four-bladed impeller which rotates in a five-bladed chamber. This is the well-known Eaton-type pump which supplies more than enough lubricant to carry away heat generated by the plain bearings in the engine. A large-capacity full flow oil filter is easily accessible for replacement, and oil mist which is present in any four-cycle engine is separated back into oil and air by an anti-turbulence chamber. This air is sucked back into the inlet manifolds and burned with the incoming fuel/air charge.

The traditional automotive-type single plate dry clutch of earlier BMWs is retained and is relatively light and very smooth in operation. Wheelies weren't possible when shifting into second gear at the drag strip on the R60/5, however, and we got quite a bit of clutch slip after the third quarter-mile run and an accompanying increase in clutch lever free play. On the way home from the strip the clutch cooled and the adjustment returned to normal. The clutch on the R75/5 we last tested seemed much more resistant to heat and slippage.

All BMW /5 series machines have lighter flywheels than their predecessors, which makes shifting an easier task. Waiting for the revs to unwind between shifts to permit the engine/

BMW R60/5

SPECIFICATIONS

List price	$1808
Suspension, front	telescopic fork
Suspension, rear	swinging arm
Tire, front	3.25-19
Tire, rear	4.00-18
Brake, front, diameter x width, in.	7.9 x 1.2
Brake, rear, diameter x width, in.	7.9 x 1.2
Total brake swept area, sq. in.	59.5
Brake loading, lb./sq. in. (at test weight)	10
Engine, type	ohv opposed Twin
Bore x stroke, in., mm	3.07 x 2.9, 73.5 x 70.6
Piston displacement, cu. in., cc	36.6, 599
Compression ratio	9.2:1
Claimed bhp @ rpm	46 @ 6600
Claimed torque @ rpm, lb.-ft.,	35.5 @ 5000
Carburetion	(2) 26mm Bing concentric
Ignition	coil and battery
Oil system	gear pump, wet sump
Oil capacity, pt.	4.75
Fuel capacity, U.S. gal.	5.0
Recommended fuel	premium
Starting system	electric, kick
Lighting system	12V alternator
Air filtration	treated paper element
Clutch	single-plate, dry
Primary drive	none
Final drive	(3.36) shaft and bevel gear
Gear ratios, overall:1	
5th	none
4th	5.09
3rd	6.31
2nd	8.66
1st	13.09
Wheelbase, in.	55
Seat height, in.	33
Seat width, in.	11
Handlebar width, in.	30
Footpeg height, in.	11.7
Ground clearance, in.	8.2
Curb weight (w/half-tank fuel), lb.	440
Weight bias, front/rear, percent	44/56
Test weight (fuel and rider), lb.	590
Mileage at completion of test	1150

TEST CONDITIONS

Air temperature, degrees F	79
Humidity, percent	42
Barometric pressure, in. hg.	30.04
Altitude above mean sea level, ft.	350
Wind velocity, mph	8-12
Strip alignment, relative wind:	

PERFORMANCE

Top speed (actual @ 6900 rpm), mph	102.70
Computed top speed in gears (@ 7000 rpm), mph:	
5th	none
4th	104
3rd	84
2nd	61
1st	40
Mph/1000 rpm, top gear	14.9
Engine revolutions/mile, top gear	4040
Piston speed (@ 7000 rpm), ft./min.	3195
Lb./hp (test wt.)	12.8
Fuel consumption, mpg	50
Speedometer error:	
50 mph indicated, actually	45
60 mph indicated, actually	56
70 mph indicated, actually	64
Braking distance:	
from 30 mph, ft.	22
from 60 mph, ft.	124
Acceleration, zero to:	
30 mph, sec.	2.9
40 mph, sec.	3.8
50 mph, sec.	4.7
60 mph, sec.	5.9
70 mph, sec.	9.2
80 mph, sec.	13.1
90 mph, sec.	19.5
100 mph, sec.	29.2
Standing one-eighth mile, sec.	9.22
terminal speed, mph	71.51
Standing one-quarter mile, sec.	15.57
terminal speed, mph	83.95

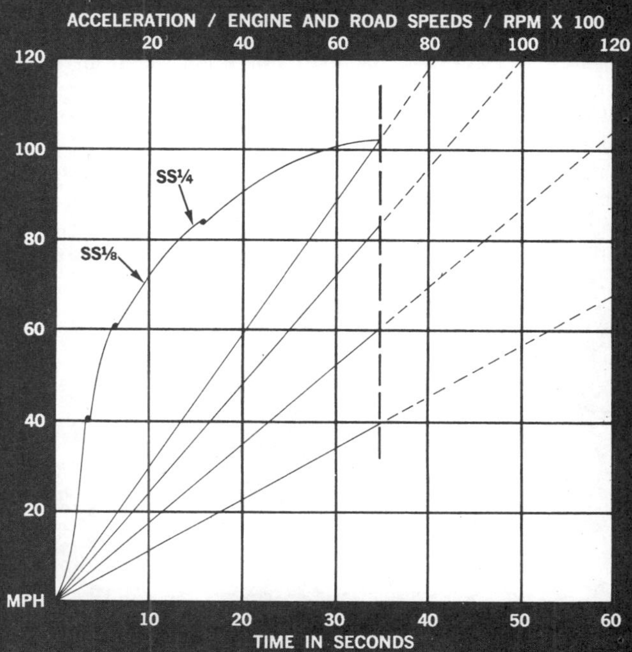

transmission speeds to synchronize was annoyingly long; and with a non-synchronized transmission and a clutch that turns at engine speed, this pause was necessary to avoid the dreaded *clunk* when shifting. The *clunk* has been reduced in intensity, and with care in matching the engine/road speeds, practically noiseless shifts can be made. Gear width is substantially greater than in earlier models, assuring long, trouble-free life.

The remainder of the drive train has been synonymous with BMW since the year one. A shaft drive system is preferred by many serious long distance road riders because of its freedom from adjustment and inherent cleanliness. The shaft itself is neatly ensconced within the right hand swinging arm member. Up and down movement of the drive shaft is allowed by an automotive-type universal joint and the driveshaft is splined to permit the slight changes in wheelbase which are attendant to the swinging arm's up and down motions. A hypoid gear in an oil bath translates the longitudional circulatory movement of the driveshaft to the rear wheel.

The wheels show more evidence of BMW's weight reduction plan. Light alloy rims are used and feature cross indentations in the central portions which prevent the tire's bead from slipping into the bottom of the rim if there is a puncture or blowout. Large cast aluminum brake drums with steel liners feature brake linings that were originally developed for the latest (ca. 1963) drum braked Porsche automobiles which are highly resistant to fade. The latest redesign of the BMW's wheels makes use of narrower brake shoes but with a larger drum diameter.

Not surprisingly, the R60/5 stopped in very nearly the same distance as the R75/5 we last tested, and even when packing double through the mountains where one uses the brakes quite a bit on downhill sections, fade was minimal. Short, straight spokes are chrome-plated and add a touch of class. The straight line pull of the spokes makes for a very strong wheel assembly.

A rather unique method of actuating the front brake's double brake cam pivot arms is employed. Instead of using an adjustable rod between the brake pivot arms, the single brake cable stretches between the two, and when the front brake lever is pulled the arms move toward each other, applying equal pressure to each arm, or at least in theory—a clever and simple method.

Large diameter oval section tubing is used in sections of the duplex cradle type frame, and the rear member which serves as a support for the seat and fiberglass rear fender is detachable. Noted for its rigidity, the frame permits a slight, controlled

flexing in the longitudinal plane, which seems to give the best handling characteristics. Hefty mounting bosses support the engine/transmission package, which is bolted directly to them without benefit of rubber mounts.

Last year we complained slightly about the seemingly wasted movement in the front forks. With a potential of nearly 8 in. of travel, we felt that the forks springs were too soft, allowing too much compression travel to be wasted. Evidently the engineers at BMW agreed, because the 1972 models feature longer and slightly stronger front fork springs. Some minor changes to internal bushing length and rebound damping have also been made.

Whether riding solo or packing a passenger, the front forks are quite good in terms of ride comfort and keeping the front wheel where it should be—in touch with the ground. Even crossing railroad tracks at speed showed little tendency for the forks to bottom out completely, and the enormous reserve of upward movement available wouldn't allow them to top out, either.

The rear suspension units are well up to the job of smoothing out bumps in the road. A large handle at the bottom of each unit permits the shock absorber spring to be adjusted for increased loading imposed by baggage or a passenger without taking the toolkit apart in search of the proper spanner. Damping is a bit on the soft side, which allows the wheel to follow undulations in the road, and isn't quite firm enough for spirited cornering over rough surfaces. With two aboard and the shock springs set in the middle position, the ride was somewhat akin to that of a Cadillac: soft, but not really mushy. After all, the R60/5 is a touring bike, not a racer.

Starting the BMW in the morning is aided in no small way by built-in accelerator pumps. This device is located in the carburetor's mixing tube and is activated automatically by opening the throttle. First the fuel level in the carburetors can be raised by depressing the float plungers, then a couple of blips of the throttle further enriches the fuel/air mixture. There is no choke. Other than a slight amount of leaking after they had been tickled, we had no complaints about the Bing concentric carburetors. Almost perfect adjustment by the shop personnel at the distributors, Butler & Smith, Inc., also aided in obtaining an excellent consumption figure of 50 mpg under normal riding conditions.

Lighting and associated wiring must also be termed excellent. Especially appreciated was the headlight's beam pattern. On low beam a rather sharp cut-off to the left helps keep from blinding oncoming traffic, while on high beam the swath of light is wide and full. Large and bright is the taillight and the turn indicators are easily seen by other traffic.

It would be easy to go on commenting about the BMW R60/5. We could say how much we like the idea of easy accessibility to parts which need adjustment, as well as the extended service intervals. Or we could complain that the peculiar torque reaction you get in the lateral plane when blipping the throttle was at first disturbing, as was the phenomenon of the rear end's rising under hard acceleration and squatting under hard deceleration.

But it's easy to say just how much we appreciate the refinement and quiet operation of this motorcycle. In terms of lack of noise, both exhaust and mechanical (BMW claims a maximum of 75 db at full throttle), and the attention to detail found everywhere, the BMW is tops. If you've outgrown the penchant for seeing how many old ladies you can frighten with your noise, are tired of constantly cleaning an oil-leaking motorcycle and want to ride and be sure you'll get there, try a BMW. Chances are you'll like it!

She was a phantom of delight
When first she gleam'd upon my sight;
A lovely apparition sent
To be a moment's ornament: *W.W.*

BMW R75/5

TEST: John McDermott
PHOTOGRAPHY: Vic Barnes
HEADING AND CAPTIONS: William Wordsworth

THERE ARE THOSE WHO would have you believe that £1300 is a lot of money to pay for a motorcycle, even the best motorcycle imaginable. And there is a company known as the Bayerische Motoren Werke who imagine they have built the best motorcycle ever. The price tag on the handlebar reads £1298.

The organization is impeccable. You telephone for a test 'bike, dates are suggested and rejected until a mutually acceptable date is arrived at, and you make a ring round the corner of the desk diary.

Then one morning Terry Wogan blarneys his way into your life from somewhere out there in the darkness of the semiconcious half-world and tells you the time, date and weather. The date means something. You work on it. Then, before the buzzer shatters your head clear of dreams, you leap out of bed, grateful that the Met Office has arranged only intermittent showers, in the full realization that today is The Day.

Shreddies

BEING LOANED A NEW R75/5 is something to look forward to. It dominates one section of the brainbox over most else. You fumble into the Shreddies pack and spurt milk and sugar over your newspaper. Your taste buds are geared to receive toast and marmalade and stunning confusion results before you realize that you picked apricot jam off the shelf. You even forget to listen for the morning's deliberate error, and make a few yourself. Like odd socks, only noticed when you've reached the bus stop. Dammit. Home again to find that in your eagerness to get to that big motorcycle you forgot to pick up your house keys and the only open window is that of the first floor bathroom.

With great labour you arrive at BMW's office. The civility is almost unbearable. You get a coffee that takes half a day to cool. Jeez, don't they know you just risked your life jumping off a bus at Hammersmith and catching another on the move just to be here as soon as possible? And what kind of people are these? There must be eight BMWs out there and they want to drink coffee?

You learn that your 'bike is being refuelled, having the tyre pressures checked and the dipstick dipped, your coffee cools and then a phone call says everything's ready. You're almost there. Now all you have to do is get out of that office without being rude, not too fast. Thank you Alan, See you soon Doug. 'Bye Alberto. You back up to the door. Thanks for the coffee Glyn. Thank you, thank you, thank you. A mechanic runs over the machine for you. One more time. Thank you Dave. And then everything has been said and you're on your way.

Showers

MAYBE I WAS TRAVELLING too quickly. Maybe half the population of London is deaf to anything that emits less than 80 decibels, but less than four miles separated me from that half finished coffee before a whole squad of kamakazi pedestrians had all but thrown themselves under my front Metzeler. Most of them unaware of my presence even though I was the only vehicle on the street. Now I can't take too much of that. I'm a sensitive being who will damn near plough into a tree to avoid a road running rabbit. Mesmerized moths don't bother me much, but people and bunnies do. So a trip out of town and away from the suicidal hoards seemed in order. Super-quiet motorcycles present problems, it seems.

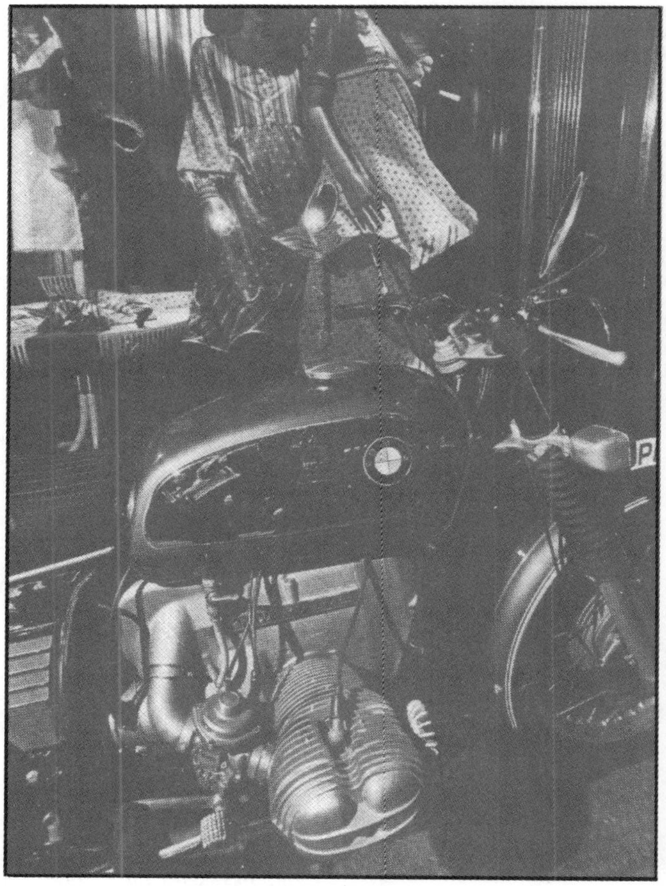

A dancing shape, an image gay,
To haunt, to startle, and waylay.
W.W.

I travelled south toward the A24 and then changed my mind. The A29 could be picked up east of Guildford and the road to Bognor Regis is a twistier, narrower, altogether nicer road than the Brighton road, especially if one is wanting to put a two wheeler to the test and find a few shortcomings.

As I reached Epsom the sky opened and the Met Office boffins must have smiled to themselves. It was the first of their promised showers and it made the going a bit uncertain. I moved nearer the crown of the road to avoid the oil slick that runs down each side of every road in the country. The motor car seems to have donated more than its fair share to the Epsom Borough Council. The tyres were no more than sixty miles old, and unscuffed, not the best condition for tyres that were to take me on a 130 mile test ride, but they did nothing silly. They skated a mite whilst changing lanes on roundabouts and slipped a bit on fast left handers, (why can't I take right handers at the same speed I can left handers?) just enough to make me feel like I was a really capable rider. The rain became more than just a shower but I didn't mind. If you really want to find out how good/bad a 'bike is, wait for the first rain in a week and then get out there and ride fast.

I soon had the controls weighed up. Clutch and brake lever and twistgrip were where they always are and the footworks were the right way round too; brake on right, gear-shift on left. You disagree? The British-built motorcycle seems to be the only vehicle in the world with a left foot operated brake pedal, and *that* seems odd to me. The right hand thumbswitch was for right and left turns and for starting. The left thumb was used for dip, flash and honk. Although I was sure I felt the machine make a half attempt at steering off right at tortoise pace in traffic, out on country roads it was perfect. Well balanced, responsive and absolutely vibration free.

Snap

MANY RIDERS CONSIDER appendages such as turn signals to be fatuous. I always did, but was soon using them to the full. In drizzle, the fitful orange glow gave oncoming traffic no excuse for not seeing me overtake my slower southbound companions. They were there to be used, so I used them.

The rain stopped suddenly. Snap. Just like that. I could see the join mark between wet and dry as I passed from rain into sunshine. Those Met boys must have been pulling switches. I checked speedo and watch to find I had covered 37 miles in 34 minutes, not a bad average in bad conditions, especially on the A29 twixt Beare Green and Pulborough where overtaking is restricted and an abundance of speed limits prevail. That covered BMW handling. The 'bike had no foibles. No quirks. Lay her down and feed in the power, or, on wet roads, lay her down a shade less, use the knee-dangle method and feed in the power. There's an awful lot of poke to be drawn out of that funny looking motor, so be sensible. I tried a very fast sweeping left at a rather naughty speed and very quickly learned that I was perhaps stretching my luck. I rolled it off and sat up a bit. She wiggled her hips, but not enough to frighten me. I cussed inwardly for asking too much of both 'bike and self and made mental note of gear and speed, filing the information for my next crack at that bend and decided to be more professional about it next time.

The suspension was set on mid position when I collected the 'bike and I left it there. Although I was mentally geared for testing and not just riding, that is using all my faculties for noting all manner of things for regurgitation later, when paper was fed into typewriter, the one thing that escaped me completely was the suspension. Usually one has cause for remembering one or two mannerisms, like pattering or chopping, wallowy motions or hobbyhorsing, or poor damping and restricted spring travel. Or something. But I missed everything, so I can only assume that it was near perfect. I did, however, notice all eight inches of fork travel when someone in a Rover came at me, crossing a double line to do so, and I braked accordingly from 70 odd to near zero in a matter of yards. The black rubber bellows gasped a bit and I wasn't slow in doing likewise. That brought my attention to the braking department.

Stoppers

BOTH DRUMS ARE 7.8 inches in diameter, giving 66 sq.in total braking surface, twin leading shoe front, single on the rear. Finned alloy jobs they are, with chromed sideplates to tart them up a bit. They work extremely well, the front more so than the rear, as the foot pedal needed quite a lot of pressure to match the mildest grip on the front's lever. But if you are a piano player, or one of those people who can make circles one way over your head with one hand, and contra circles in front of your tum with the other, you'll have no problems of applying two pressures and attaining a common effort from each brake drum. The front shoes needed chamfering to allay their desire to work pluperfectly at slow speed, a situation that could lead to postponement of your wedding or a place in the Luton Girls' Choir. They are good stoppers and no doubt about it.

Comfort rates highly too. I'm a big lad, standing six foot three in my vest, and I have size eleven feet (to boot). Once in a while I found the carb induction pipes and the bottom of the carbs coming down to meet the topsides of my feet. Usually on bumpy roads when my boots were inclined to be bounced off the pegs, sometimes it happened when I used the front brake to good effect. A small criticism, and really I can't blame BMW for the size of my plates, but a couple of inches more tootsy room would be appreciated. Everything else about the 'bike fitted well, 'bars the right width, tank the right shape, saddle the right height. Made to measure. But for shorties there may be problems. Seat height is an amazing 33½ inches, so legs shorter than Ann-Margaret's might cause problems.

Schnozzle

EVERY MACHINE EVER made has its shortcomings. As far as I could ascertain, the BMW has but one. Yes, altogether now . . . The Clunky Gearchange. Oh, how annoying it is to eventually get the 'bike/rider relationship just as it should be only to find there is one thing you just can't get together on. When the engine was cold I could stomp up and down on the gearshift with nary a muted thump from the 'box, but when she warmed up, she insisted upon telling me, and, it seemed, everyone within the parish boundary, that another ratio was being sought. Such is the penalty for having an engine speed clutch.

Later in our ten day aquaintanceship we came to an agreement. I had it my way on most changes up and down except those from third to second on the way down, although on one occasion she forgot that she was supposed to make a noise between those two gogs. I tried again to lose (whereas Jimmy Durante was after finding) that accursed chord. And like Schnozzle, all I found was failure. What a catastrascope.

The ratios were nicely suited to both town and country riding and if one could forget the clonk (but one couldn't) there were no other complaints.

Quite a lot of work has gone ito making the R75/5 more of a 750 than she has been heretofore. Before now she has been thought of as a gentleman's 750 – not very fast, not particularly quick either, a fairly capable 650 perhaps. No more, brother. Things have happened. Like the kilo shaved off the flywheels to give better acceleration. Zonk. Zero to sixty takes no more than six seconds and max comes up as quickly as most other seven point fives. Claimed speeds in each gear at 6200 rpm are: 42; 63; 86; and 109 mph.

As usual, the manufacturer is telling fibs. Only this time, they have erred on the credibility side. Indicated speed on the machine I rode was a nick over 120. Even allowing for a normal speedo error, that's a bit more than the claimed 109 mph. And a story did reach my ears that a police force timed the R75/5 at over 119 mph using electronic equipment. I'm not surprised, these big BMs are loath to be sloth.

Stalls

GETTING 95 MPH IS pretty easy meat for most 750s built these days, but getting *over* the ton isn't as easy as most coffee bar bigmouths would have you believe. The difference between

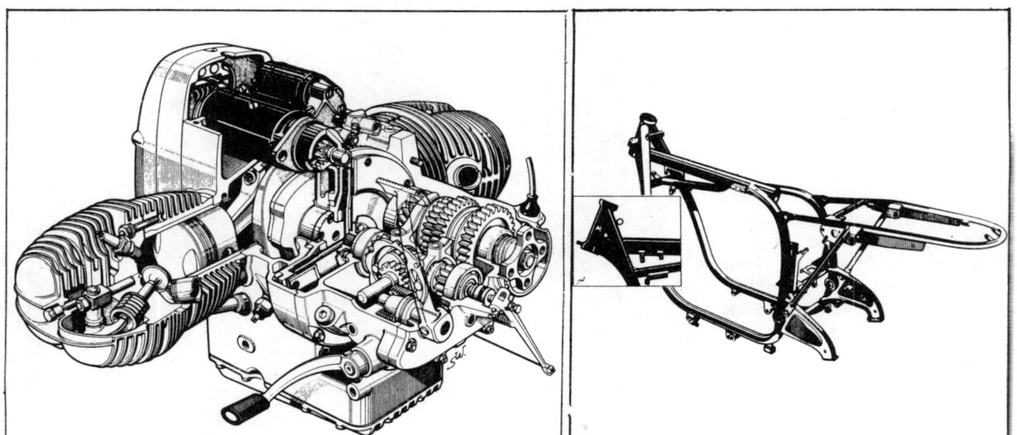

And now I see with eye serene
The very pulse of the machine;
W.W.

95 mph and 105 mph is enormous. A true 110 can be as hard to find as tickets for a Presley Personal Appearance. Keeping up a high average speed is as difficult as getting seats in the centre stalls. BMWs are built for high average speeds. Whilst cruising an R75 at an indicated 115 mph on a continental autoroute last year I came upon a bunch of Commando mounted Frenchmen. They gave chase and stayed with me for about six miles only to fall back in wise respect for their motors. I was on stage, they weren't even in the foyer. That afternoon I covered 210 miles in 2 hours 42 minutes including a short stop for fuel and a stretch. And every mile was enjoyable, full of confidence – cocksure even, with not a cloud in the sky to dampen my spirits. But that was long ago.

The ten days I had the new test 'bike gave me an opportunity of getting together with it for a mere 700 miles. Had I not been deskbound for most of that time I would have treated myself to something nearer 2000 and doubtless enjoyed every furlong. I took her out on all kinds of run; the daily frustration of commuting, a few wet seaside trips, odd business calls of varying mileage and countless hops to the off-licence. The overall fuel consumption was 56 mpg. On long runs she averaged 47–64 mpg depending upon my mood and desire for rapidity. Oil consumption was nil, a great pity as my wife is collecting sherry glasses from the local garage and needed more vouchers. In the good old days one used to pull in for juice at the next garage on the left after switching to reserve, nowadays one is tempted to hang on until the combine offering the most trinkets looms up ahead. Which brings us to fueltanks.

Stamps

IT IS PRETTY GENERAL that Quad stamps or free tights or moving pictures of Charlie George are offered for four gallons or more. And it is equally general that motorcycle fueltanks hold three and a bit gallons. This year's stock BMW is no exception, the tank has a bare 3.8 gallon thirst when it comes to get-off-and-push. Refuelling immediately after turning the petcock to the 'R' position, leaves you with a bill for 3½ gallons and a ribbon of single stamps. It would be a nice gesture if manufacturers made four gallon tanks standard and added a bit for reserve. It would allow us to presumably spend more time getting where we are going, less time planning on pulling into forecourts, and give us the little bonuses that car drivers enjoy. One other thing that car drivers benefit from is the shape of airline-to-valve equipment. If you have ever had to struggle to attach an airline to a 'bike's tyre valve by threading the damned thing through spokes and around hot silencers with all the restrictions caused by large diameter brake drums, suspension struts and mudguards, you know just what I mean. Laverda owners more than most. With the BM though, there was no problem in any of the many garages I called at. The glass fibre mudguard on the rear, although deeply valanced, sat high enough from the upswept silencers to allow easy access to the valve. It wasn't even necessary to slacken the security nut and bend the valve outward. And the same goes for the front, helped considerably by the lack of bracketry holding the blade to the fork.

The same frame is used on all three BMW models. It is of single backbone, double cradle type using welded tube, and all load carrying members are of conical or oval tubing. Bolted into this frame, and hanging out over both sides is the engine. Easily the largest and tidiest 750cc motorcycle engine ever sculpted. Everything that has to be appended to the works is tucked inside the smooth alloy cases. The two cylinders, slightly offset, stick out the sides. In front of them lies the alternator and a load more electrical gubbins, behind them lies the clutch, and gearbox. And just to square things off, the space above the clutch is taken by the starter motor whilst the bit over the gearbox is occupied by a large micronic aircleaner which ducts clean air to the two vaccum carburettors.

A countenance in which did meet
Sweet records, promises as sweet,
W.W.

Spiel

750CC BMWs ARE NOTHING new. The first was made in 1928, a side valve job with handchange gearshift, and pushing out 18 bhp at a leisurely 3400 rpm. It was quite a tool and full of innovation such as handlebar levers that pivoted at the handlebar ends, a

SPECIFICATION
BMW R 75/5

Engine: Output 50 bhp (DIN) at 6200 rpm, 57 bhp (SAE) at 6400 rpm, max. torque 43.4 ft/lb at 5000 rpm; bore 82 mm, stroke 70.6 mm, capacity 745 cc (45.46 cu. in.); compression ratio 9.0:1; constant depression carburettors, diameter 32 mm.

Transmission: Gearbox ratios 1st 2.835:1, 2nd 1.875:1, 3rd 1.364:1, 4th 1.091:1, layshaft reduction 1.375:1, final drive ratio 3.2:1.

Tyres: Front 3.25 S 19 on 1.85 B×19 rim;
Rear 4.00 S 18 on 2.15 B×18 rim.

Brakes: Full width alloy hubs, friction area 2 × 16.55 sq. in.

Dimensions: Length 82.7", width 29.1", height 40.9", seat height 33.5", wheelbase 54.5".

Weight: In road trim 452 lb with full tank; permitted gross weight 877 lb.

Electrical equipment: 12 volt system, 180/200 Watt alternator, 6.3" headlight with asymmetric dipped beam and 45/40 Watt bulb, starter 0.5 bhp (DIN).

Fuel tank capacity: 3.85 Imp. gal/4.6 US gal. including 0.55 Imp gal/0.66 US gal. reserve.

Fuel consumption: 64 Imp/55 US mpg super grade fuel at a steady 68 mph.

Top speed: 109 mph.

Acceleration: 0–60 mph: 6.2 sec.

Price: £1298 incl. p.tax.

BMW R75/5

The reason firm, the temperate will,
Endurance, foresight, strength, and skill;
W.W.

leaf spring giving suspension to the front wheel and large saddle springs protecting the rider's rump from the potholes of the late'20s.

The R75 was first introduced in mid-war. Many a wise young infantryman volunteered for motorcycle duty aboard the R75 combination, a very up-to-date 750 ohv motorcycle boasting telescopic fork and a whole host of military accessories. Gearchange was still by hand, but stylishly embedded underneath the right knee grip on the fueltank.

The /5 series was introduced in 1969 to meet the demand for a bigger model for offerance to the American and European markets. Quite a few things have changed within the crankcases since its introduction, and each alteration has made the R75 that much better. The styling for '72 has come in for much criticism, so it will be interesting to see what happens when the drapes are pulled off the '73 models. Maybe we'll even find out if those rumours about bigger engines is more than bierkeller talk.

I pulled up at a traffic light alongside a Bentley Corniche that had just glided out from the Savoy. An expensive suit moved forward and a window slid down, and whilst we waited for the green a Bermudan tanned middle-aged man asked questions about the 'bike. I gave him all the spiel that those few moments would allow. He even stopped sucking on his Habana, and when we moved off he drank in every line of the machine, D'you know, I think he was a bit envious.

I should have asked him if he thought £1300 was too much to pay for the best motorcycle imaginable. ■

Baby of the family

LONG after the R75/5 and the R60/5 had raided these shores and scooped all the glory came, almost unseen, the 500 R50/5, the baby of the family, aimed, we imagine, at a very different sector of the market from that identified with the higher performance, bigger /5s. On the face of it the 498 c.c. model is much more the "traditional" BMW, sacrificing performance as it does in favour of smoothness and flexibility. But, in fact, with our test 500 things did not work out quite like that. The R50 was no smoother than the bigger machines and, if anything, was less flexible. It develops a maximum torque of 29ft/lb at 5,000 r.p.m.; and the R75 develops 43ft/lb at the same engine speed.

The R50/5 was certainly like no other 500 BMW we have tried. On the "softer" models of old one would whoofle along in top, disdaining excessive revs. The R50/5 needed to be driven all the time. Below 4,000 r.p.m. there was very little real response; to make anything like rapid progress the tachometer needle had to be kept hovering around the 6,000 mark, which was hard work and left us with the question, "Why make an R50/5 ?". Is it produced merely to offer a (compara-

AN EVALUATION OF THE 500 BMW WITH AND WITHOUT ELECTRIC STARTER

tively) cheap bike in the range ? It sells at £947, without electric starter, some £115 less than the starter-equipped R60/5; have the optional starter (at £28 extra) on the 500 and the difference is reduced to £84. What does one get for this money on the 600 ? The cycle parts on all three machines are identical and, unless there is some secret that we have missed, all one appears to get on the R60/5 are bigger pistons and matching cylinder bores. For £84 ? Phew !

What other reason is there for producing the R50/5? To keep the traditional BMW 500 in the range, perhaps, or to woo the older rider who wants to own a BMW but doesn't want too much in the way of performance ?

All this, of course, has nothing to do with our road-test machine . . . or has it ? It came as a small disappointment to find that we actually had to *kick* start the bike. A couple of years on electric starters has completely spoiled us but the 500 did present a chance to see if the 20 lb removed from the top of the engine did anything for the handling. Our R50 came in the good old traditional black. It would have been more imposing with the large tank (the concessionaires tell us that 80 per cent of future models imported will have the larger tank). It was not so much tank size that spoiled the looks for me but the chrome panels on the side. They just aren't "BMW". Choice of colour is a very personal thing and I am bound to say that I tend to prefer the new brighter colours that BMW have introduced with the 1972 range; the deep blue is particularly attractive, I think.

The BMW has probably had more column inches written about it in this country in the past couple of years than any other motorcycle, which is due in large part to BMW (GB) Ltd being more than willing to make machines readily available for comment. As a result one not only reads about BMWs in road-test reports but finds that BMWs are being used on all manner of incidental stories. Interested people have long since discovered that if a motorcycle is needed quickly chances are that BMW will supply one without fuss or bother. Which contrasts strongly with some of the reactions we get when we need a machine from other manufacturers/concessionaires.

You will need no reminding that the BMW—all BMWs—has a horizontally opposed flat-twin engine, etc. We shall content ourselves with details applicable only to the R50/5. It has a similar stroke measurement (70.6mm) to those in the

BMW R50/5

The 500 *BMW with "small"* 1972 *fuel tank; compare it with the model shown in the photograph on page* 336

larger machines, its smaller, 498 c.c. capacity being obtained with a stroke of 67mm. The compression ratio, at 8.6 to 1, is slightly lower than for the rest of the range but considerably higher than BMW have ever had before on a road-going 500. Maximum b.h.p. is 36 at 6,600 r.p.m. Carburettor choke size is 26mm, as on the 600. The only other obvious difference is in the rear-drive ratio, which is 9/32, giving a reduction of 1 : 3.56. This compares with 1 : 3.36 for the R60 and 1 : 2.91 for the R75. All other technical details are identical on the other machines, except for weight, which is quoted as 408 lb dry (we made it 412 lb with petrol). The absence of an electric starter gives the R50 a listed advantage of 11 lb, which is strange if the starter really does weigh 20 lb, as we have been told. Claimed top speed of the 500 is 98 m.p.h.—unusually optimistic for BMW for they rarely claim that a machine will do something that it doesn't. We tried two 500s and both were just about willing to reach 90 m.p.h. with the rider flat on the tank.

The modest performance of the R50/5 set us thinking. I owned for a number of years a 1953 R51/3—that was the telescopic fork, plunger rear frame predecessor to the R50, which came before the R50/5. Its top speed, if memory isn't playing too many tricks, was at least as high as the latest model's and the motor, with its low-compression ratio and soft cams, was much sweeter. We raided the archives of BMW enthusiast Keith Sanders and came up with the following figures. They cover most of the 500s produced by BMW since 1937:

500 FACTS AND FIGURES

Model	Tested by	Date	Top speed, m.p.h.	S/s ¼-mile, m.p.h. (terminal speed)	Power output b.h.p.	Weight lb	Compression ratio
R5	*Motor Cycling*	26.1.38	92 at 6,000 r.p.m.	51.1	24 (continuous rating)	364	6.7
R51/2	*Motor Cycle*	30.11.50	88	65	24 at 5,800 r.p.m.	396	6.3
R51/3	*Motor Cycling*	31.5.51	86 (mean) at 5,490 r.p.m.	50.56	24 at 5,800 r.p.m.	400	6.3
R50	*Motor Cycle*	7.7.55	90 (mean) 94 (one way)	76	26	413 (a misprint, we suspect. It should be 430.)	6.8 to 1
R50/5	Maker's catalogue	—	98 (claimed)	—	32 (DIN) 36 (SAE)	408	8.6

Figures do not, of course, tell all the story, but they appear to suggest that, from a performance point of view, the 500 c.c. BMW has changed very little since 1938. The obvious conclusion is that the latest /5 500 needs another 8 b.h.p. to push along its extra 44 lb in adequate style!

One area in which no one is going to dispute the superiority of the present-day 500 is handling. The original R5 had a solid rear end and telescopic front forks; plunger rear suspension was added for the

/2 and /3 models, and the R50 was the first to enjoy the benefit of pivoted-fork rear springing and Earles type front forks. All were reasonable for their era but none could be considered as in the class of the present-day model. We would venture to suggest that *no* honest medium-capacity touring machine offers better steering and road-holding than the R50/5. It really is superb. The modest performance from the engine ensures there is no danger of the machine misbehaving.

What of the missing electric starter? Does it make a significant difference to handling? We are hesitant about this, suspecting that it does, but we are reluctant to commit ourselves. After all we were *expecting* it to, and this in itself might have clouded our judgement. The acid test came when we returned the 500 and were given the chance to enjoy a 750 for a week. Stepping straight off one and on to the other gave us a direct comparison, and *then* we could tell the difference. It was not, to be fair, much of a difference but where it showed was in riding fairly quickly through an S bend, lifting the machine up and putting it down again in rapid succession. The R50/5 felt lighter and more responsive. Marginally but, just about, noticeably. How much of this was really in the mind? You see, to all intents and purposes the R50/5 and the R75/5 are, when you are sitting still on one, identical bikes. One has to be particularly perceptive to glance down and say that the cylinders sticking out the side are those of the 750, say. Yet as we rode the 500 there was the very strong feeling that it was a smaller, lighter model. But, certainly, a great deal of this sort of feeling *is* in the mind.

While we are making comparisons with the R75/5, it is interesting to note that both machines returned the same fuel consumption, 58 m.p.g., over the same type of going. It makes sense, of course, for although the R75/5 has the larger (32mm) constant vacuum carburettors, the R50/5 was having to work harder for any given speed. For example, the smaller machine would do 70 m.p.h. at 5,000 r.p.m. and the larger 78 m.p.h. (these are speedometer readings; the graph published in the handbook agrees with our first figure but suggests that the higher figure ought to be 85 m.p.h. We suspect that our tachometer was a little slow). A commonly held theory is that the R50/5 ought to last for ever as it shares the same bottom end with the larger machines and therefore obviously has a greater built-in safety margin. We wonder if the reverse might indeed be the case—that the larger-capacity machines will ultimately have the longer life because they have to rev less highly to achieve a given speed. The flaw in this argument is, of course, that one is apt to ride the 750 more quickly, in any case, than the 500.

The comparison between machines leads one to think that perhaps the R50/5 is a mite "over-engineered". This, in a way, is a corollary to the argument about the long life span of the 500. But is the whole thing purely a hypothetical argument, anyway, when the design is said to be good for 100,000 miles?

The trouble with BMWs is that they are so good that one tends to become absorbed with matters such as the above because, if we kept strictly within our terms of reference, the article would merely say that the R50/5 is a superbly made, reliable, good-looking machine that will do 90 m.p.h. without too much difficulty. It is more than that. It shares with the rest of the range so many features that are desirable. Lights as good as the best. Controls that *are* the best, with neat rubber sleeves over all cable ends to keep the water out. The switches are exactly right, identical on either side, the left one to operate the horn by pushing the whole unit and dipswitch, with main beam in the logical "up" position and the headlamp flasher actuated by over-riding the dip position on the switch. On the right, the indicators are operated by the positive up and down switch and the starter (if it has one!) by pressing the whole unit. Delightful. Less so is the master light/ignition key. It is a unit that has functioned perfectly well for years on earlier models but now, somehow, BMW have contrived to make it twice as difficult to get the key in or out. Why?

Then there is the smooth, quiet engine. Less quiet, due to its all-alloy barrels, than was the pre /5 model but still quieter than anything else around. The exhaust note has not improved only because anything quieter than the old one would have to be *silent*! How do you describe smoothness? There are many excellent, smooth engines around today, some as smooth, maybe, as the traditional flat twin. None, we venture to suggest, is more smooth. Here's an example. We visited a friend 30 miles away and took a bottle of home-made wine. It was wrapped in our leather jeans and put in the tank bag. At the end of the journey

"Electrics" on BMWs are preserved in pristine condition for years in the compartment between the front down tubes

BMW R50/5

the sediment (we are not yet experts on home-made wine !) was still undisturbed in the bottom.

The motor was, as we would expect, oil tight but the rubber gaiter at the gearbox end of the drive shaft was either damaged or was a bad fit, for there was oil escaping from that area. The standard of the paint work and chrome was excellent—no suggestion of rust anywhere. But no machine is faultless. The prop stand, nominally the not very attractive fly-back design, was so stiff it would not fly anywhere. It also held the machine too upright for our peace of mind. The centre stand was fairly easy to use but one had to ensure that the machine was not facing downhill, when it would be in danger of running forward off the stand. Rear wheel removal *could* be a chore. It took, literally, seconds to free the wheel but as the rear mudguard did not hinge one then had to tilt the machine to one side to complete the operation. It was easy with assistance, difficult otherwise.

The BMW twin-leading-shoe brake was first introduced in 1954, which has given them a fair while to get it right. They have. Our brake was as effective a front brake as it has been our good fortune to use and, with a useful rear one, the stopping power of the BMW was outstanding. We are bound to report, though, that some BMW front brakes grab when damp, and BMW ought to look very hard at this problem. It would be a pity if such an outstanding brake became disliked because of this fault.

The BMW R50/5 is something of an enigma. It has most of the desirable features that one normally associates with a BMW and is sold at what could almost be termed a competitive price. But . . . the limited performance from the 500 c.c. engine must severely restrict the sales of this particular model. Although one apparently gets little extra for the R60/5's price we would consider it money well spent. The R50/5 may well appeal to the older rider who wants BMW quality but is not particularly concerned with performance. We would sum up by saying that the R50/5 is a first-rate machine by anyone's standards but by BMW's standards it could do with just a little extra to justify its existence. One thing we must say. It is almost worth buying for the tool kit ! Quality through and through . . . it is as well that the makers provide a lockable dualseat to protect the kit and the excellent pump.

A pleasure reserved for the man who has laid out £1,000 . . .

WE HAVE often advocated the advantages of riding more than one example of a test model to get, so to speak, a second opinion. It is not very often that we have the opportunity. This was one of those occasions when we did for, after the R50 test went to press last month, BMW asked if we would mind helping to run in a new R50/5. Hands up all those who would say "Frightfully sorry, we are too busy". Neither did we! As good (or bad!) fortune would have it, the R50 test was held over from last month so we have the opportunity to write a little more on the subject of the BMW R50/5.

Our good friend Alan Aspel, now in charge of press and public relations of the motorcycle division at BMW(GB), was just a little put out when I arrived at Byfleet to collect the R50. "I *did* ask the mechanic not to ride the bike before you came but he took it around the block". The speedometer showed just two miles; Alan had planned to have it ready with a line of zeros.

There is something special about being the first to ride a new motorcycle. When that new motorcycle is a BMW it is extra special. The two miles didn't count as far as I was concerned. For the next few hundred miles the BMW was all mine and for me alone was the pleasure usually reserved for the man who has just laid his £1,000 on the table. This R50 was so good-looking it stopped you in your tracks. It was the fresh and eye-catching Polaris silver, but again it wasn't just that that made the BM such an eyeful. Thankfully BMW(GB) have reverted to importing the majority of machines with the old type 5½-gallon tank. It will cost owners another £6 or so but when they are spending the kind of money BMWs cost that merely counts as small change. Another hidden, but welcome, extra was the electric starter. I don't care if it does weigh 20 lb and costs £28. I like it.

The BMW purred away from Byfleet like a happy cat. The engine was so smooth it was almost unnoticeable. The brakes, on the other hand, were to be positively frightening until they had bedded in. It took about 50 miles to bring them to reasonable efficiency. My journey home was just 10 miles. By that time the BM was smelling that lovely warm new smell. It was everything that the model we have tested was but . . . it was more. The pipes, in the first few hundred miles at least, had not a trace of blue. Evidence that ignition and carburation were just about spot on. It felt like it too. It was quite the best /5 BMW I have ever had the pleasure of riding. Maybe, after a few thousand, it will loosen up, become less smooth and more noisy. Maybe. Right now I don't care. I am riding, and enjoying, every minute of a delightful motorcycle. It even *feels* more powerful The limit at the moment is 4,000 r.p.m. Far too low to get any real performance figures but one cannot escape the feeling that is has found a little extra poke. Perhaps the lighter (by 1 kilo) flywheel has made a difference? Just one small criticism. The suspension is definitely harder than on previous models, Road irregularities can be felt when carrying a passenger, even with the rear suspension setting at its softest. Which reminds us, whatever happened to the gas-filled rear units that BMW claimed adjusted themselves to the rider's weight?

Tomorrow the R50 is going to someone else. I am off on holiday and the running in will be completed by the time I return. I am usually sad to lose road-test machines from my garage. In the case of the BM I am more than sad. That machine would be welcome to live there any time. You will have gathered that I liked it. B.P.

BMW (GB) have reverted — at extra cost — to importing machines with the old-type big tank

MOTOCRAT

Story and pictures: Roy Buchanan

IF YOU ARE THINKING of buying a BMW just to impress your friends, don't bother. Don't misunderstand me, it's not that they wouldn't be impressed, a BMW never fails in that way, I am merely saying it would be such a waste. I've noticed over the past few years that people seem to buy their 'bikes for affect rather than effect. I recently fell into conversation with the French owner of a Honda 250. We had an hour and a half to kill on the cross-channel ferry so we talked 'bikes (what else!) I asked him what he thought was the main difference between the French enthusiast and his English counterpart. He said "It's their attitude mainly. The Englishman treats his 'bike as a vehicle, a means of transport as well as a labour of love. He has the big fairing and the satchels (panniers)

MOTOCRAT

The R60/5 was adorned by a number of BMW-supplied extras; the motifless leathergrained tank cover, a touring screen and Pony Express type leather panniers.

whereas the Frenchman now, to him his 'bike is an instrument of his dream world. He has the noisy exhaust, the racing 'bars, the thin hard seat and he roars down to his local café and sits outside on his 'bike talking to his friends. I'll tell you something, there are hundreds of Honda 'Fours' in Paris and I'll bet you 75 per cent have never been ridden out of town.'' Do you see what I mean by a waste?

I often look at some of today's super machinery and wonder, for whom are the factories making the 'bikes? At the bottom of the list must be the touring motorcyclist and as for long-distance man, well, he passed on with gaslit streets. Look at some of the features on the current crop: Spartan mudguards, sometimes nowhere near the wheels, which leave your engine covered in filth. Short dual seats, great if you are married to a big girl. A labyrinth of exhaust pipes that look like an office block's central heating system. How about that garish creation called a Hurricane? I don't know about you, but that does not blow up any storms with me. It seems that sensible equipment is considered dull; image lacking. The makers say they are guided by our demands, but do we really want this sort of thing? My own personal acid test of a big machine is how far, how fast, and how much trouble? Under these three headings comes a multitude of points like vibration, petrol consumption, lighting, brakes and rider fatigue. Feed all this through my computer and the end result is a number of motorcycles you could count on one hand and have a finger or two to spare. One of the chosen few is our test machine, the BMW R60/5. Over the past 18 months I have had the opportunity to ride several BMs. They have ranged from the baby of the marque, the R50/5, to the R60/5, including a white police specification model, to a beautiful blue R75/5. Of them all, the last one I had was the sweetest. This black R60/5 was timely for I was to take it on a 5-day trip to Switzerland – and in November.

It was during my return journey that I would willingly have stood on a soapbox and shouted, ''long live BMW''. I left Deitingen in Switzerland at 11.15 in the morning and rode non-stop (except for petrol) to Brussels via Basle, Strasbourg, Nancy, Metz, Thionville, Luxembourg and Namur. I arrived in Brussels just before 6.45 in the evening; time taken 7½ hours; distance covered 410 miles. This little spin gave me a variety of roads and weather but I think the most memorable was the back road from Nancy to Metz. It was typically French. Straight as a die but lumpy. It would plunge down a hill then up the other side. When I reached the top there was the same thing all over again, and the next time, and the next. It made your blood tingle and your skin go pimply. My God it was superb, 85–90 was my speed as I reflected on my gratitude for shaft drives. It's not that I have a total aversion to chains, it's just that I think a shaft is so much more practical. It was on this blast through France that I thought of what a chain would be doing on such a road. Flexing and bending, stretching and snatching, wearing and tearing. Running dry of lubricant, picking up dirt and grit. Not so on the BMW. Revolving silently away submerged in oil, a shaft is, as our old editor used to say, civilised.

When I collected the R60 it had about 2500 miles on the clock so I was able to open it out. The first couple of days I was restricted to commuting. It would be a gross injustice to ride a big mount like this one round town then convict and sentence it. Being new to the

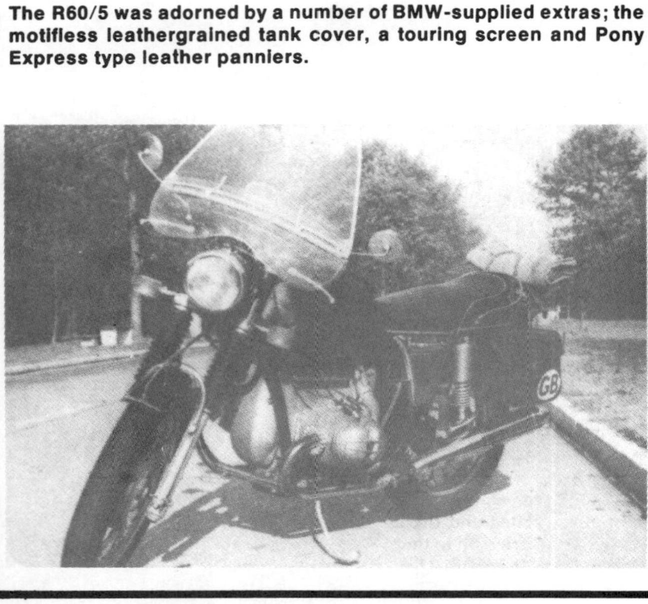

machine I found it was rather heavy in London's stop-start traffic. After I had been to the continent I discovered to my delight that the metropolis was no problem after all. It is a question of familiarity (not breeding contempt but content). I feel I should warn newcomers to a characteristic peculiar to horizontally-opposed engines which can be disconcerting. If you blip the throttle the torque produces a sideways sway on the 'bike, nothing lurid but feelable. If you are trickling along, i.e. a prospective buyer going round the block, and blip the throttle, it can be unnerving. Again, like the weight, it becomes camouflaged with use rather like the train that thunders past the back door shaking the crockery off the Welsh dresser whilst the occupier sits quietly reading the paper.

When you have a trip arranged, those last few days seem to drag as your excitement mounts. You are dying to point the thing out of town and go, then when the time arrives you wish you were not going. It was a cold, miserable Sunday night when I quietly closed the bedroom door wherein my two small sons lay deep in slumber. I knew the BMW would not waken them thanks to those huge, efficient silencers. For the last couple of years these sound trappers have been upswept to give greater clearance and to allow the condensation to run forward into the exhaust pipes where the heat steams it away. Gimmicky I thought, but if it saves the things from the demon rust I am all for it, these items are not cheap. My friend's "K" registration 750 Honda had silencers you could have strained peas through thanks to corrosion. The makers wanted £60 for new ones. He decided to fit the Dunstall arrangement, four pipes siamesing into two silencers. £28. Cheaper, but still not cheap. Must we have all these flashy pipes and boxes with their accompanying expense?

In half an hour I was cruising down the sodium-lit section of the M2, the BM was enjoying the dense night air. The engine sounded like an engine should, sweet. The motorway peters out at Canterbury, so from there it's back to the old type road. I swept almost noiselessly through the small villages with their deserted streets. I expect the inhabitants were either watching the Sunday night film or shoring up the bar in the local pub. I would be considered strange for not doing something similar, but I would rather be burbling my way down to Dover on a 'bike than getting bloodshot eyes in the saloon of the Frog and Nightgown.

There is something to be said for night riding on a BMW. The lighting is first class. The rear unit is a large visible rectangle incorporating a bright stop light, which assists the automatic pilots fitted to most cars by telling them that the motorcyclist in front is braking. The headlight is particularly effective. The main beam will illuminate the full width of the road permitting fast travel after dark. But to me one of the safest fittings in the light department are the indicators. I must confess that when these amber winkers first blossomed on a motorcycle I took a narrow minded, old-fashioned view. They were not for me but the incompetent novice who could not be relied upon to give a decent hand signal. Then I took the police R60 to Derbyshire one black Saturday night. I missed my turning off the A6 and, to avoid getting lost, a "U" turn was necessary. I was on a dual carriageway with breaks in the central island every now and then, just made for "U" turners. The first one I missed because I was too close by the time I saw it, so I continued. I think the drivers in the fast lane must have been practising for the 24 hour Le Mans. To such "aces" a lone motorcyclist looking for a gap to turn into is mere flotsam. A gloved hand signal would have been laughable but those bright flashing lights secured for me a modicum of safety. My mind had been broadened.

Meanwhile back at Dover: After sleeping the night away like a dead dog at the Brooklyn Guest House – Mrs. Brooker does a lovely B & B here for a reasonable sum – I awoke at 6.30 to the sound of pouring rain. My first thought was for the 'bike standing on the forecourt, pots sticking out, HT leads exposed to the elements. Would she start? I had taken the precaution of spraying them with WD40 before leaving home. The sound of a firing engine dispelled my anxiety. I cannot speak too authoritatively about the earlier BMWs, but the /5 series are not good starters. I have tried two methods, take your pick. At first I used to tickle one carb and after 11 attempts – well 10 or 12 anyway – with the starter I could get one cylinder to fire and eventually shame the other one into joining the fun. Later, I discovered to my – and the battery's – relief, that I could succeed quicker by blipping the throttle the split second she fired. The second method cut my starting time by half. This nonsense always occurred first thing in the morning, the rest of the day it was instant touch ignition. Real show-off stuff . . . except!

As I've said, this was not my first BMW, nor was it my first experience of the following little crippler. The morning I left Switzerland I pressed the starter and instead of turning the engine I got a sound like a Crystal Palace supporter's rattle. I spoke nicely to it but it refused to engage and start the motor. A flat battery? No, the electrical system was fine, and after a 600 mile journey the battery could hardly be flat. I had the identical symptoms on an R75 last Summer so it's not a one-off complaint. BMW, you must look into this because it puts pimples on an otherwise beautiful face. It may be significant that the starter motor is a relatively new feature and may, as yet, be a little unrefined. Now, a cry to the makers from a rider. Please don't take away the kick starter. After vainly rattling the electrical device I started her with an old-time kick. But for this I would still be yodelling for help in the Aare valley.

I was delighted to see bright skies as we slid into Calais harbour. I bid adieu to the Honda owner from Paris and took the coast road north. I am always pleased when I reach the Brussels motorway for the road up into Belgium is dull (I'll bet it was even duller in 1940). It's on a border crossing trip like this that you appreciate a nice tick-over. I was to pass through Customs no fewer than 12 times. To hammer down an autobahn at 80 for hours, then sit at the border post with the engine slowly chuffing away always endears a 'bike to me.

The BM and I fumbled our way through Brussels and sped on to Namur. I was getting excited for I knew what was coming. The delightful run from Namur to Luxembourg. It would be difficult to judge who enjoyed it most, the 'bike or me. Three lanes wide, beautifully smooth, this road is a dream, a BMW's dream. It was for this sort of going that the German heavyweight was intended. Eighty, 85, 90, coming down to 30–35 only to pass through the small towns of Marche, Bastogne and Arlon.

The City of Luxembourg, the sign told me. Any city at teatime is no place for a motorcyclist with a sense of self-preservation. You can bet your last penny the minute you take your eye off the lights to look for a sign they will change. Then as you reach for your brakes you notice you are just about to drop into a tram line. For once in my life give me the talent of Joel Robert! Big heavies are never at their best in city streets and the point is rammed home when passing through any continental town by the way you get torn to shreds by the young locals on their sporty mopeds. All this apart, I enjoyed my meander through the home town of the famous radio station.

The noise faded and the lights dimmed as I carried on to the German border at Remisch. The road twists and turns in a diabolical fashion. Good as the BM's lights are they could have had a couple of partners to give me a better feeling of security. I came up to a left hander, oops too fast, braked, leaned over and this bend simply went on and on. Nothing I could do but keep her over and feed the power on. It handled magnificently. She gripped the road and did as she was told. I liked that very much.

What I did not like was the handling under different circumstances. On long gradual bends a distinct tail wag becomes apparent. At first I was horrified, surely not on a BMW. But on the next bend there it was again. To hit a ridge or seam in the road whilst congo-ing like this, was heart stopping. I have never had this before on a Bavarian motocrat, it just had to be the panniers. They were the leather continental type which could be taken off in a minute and carried like a pair of small holdalls. I detached them and, with one in each hand, immediately listed 20 degrees to starboard. I did a quick stevedoring job on my luggage and found a remarkable improvement, but there was still a hint there. In fairness, the handling of the other BMWs I have ridden without panniers, was second to none. I had a discreet word in the ear of a BMW Club member and without hesitation he said: "It's the panniers." Over to you Herr Designer.

It was seven in the evening when I breezed into the industrial town of Saarbrucken. The central area bears the hallmarks of most town centres, big shops, bright lights and confusing one-way systems. I got lost. It amused a couple of bus-queues to see me trickle past at regular intervals with the corners of my mouth growing ever downward with each circuit. Relief came in the shape of an illuminated "Polizei" sign protruding from the corner of a building. My problems were solved by a patient, English speaking, inspector. As I left there were two officers with their heads together bending over the 'bike. Working on the principle of a look at yours for a look at mine, I took the opportunity to peep into their yard and saw, surprise, surprise, two white BMWs. Having found my haven for the night I was treated to German hospitality in a manner royal, but the details will remain my secret or I will be called bourgeois again.

The following morning I rose to find dry roads outside. How lucky can an MCI man get? The BM had been in a garage all night and started fairly quickly by the adoption of Method Two. We, the 'bike and I, burbled back through Saarbrücken and on to the French border. While still in Germany I took

MOTOCRAT

The handlebars fitted were not altogether comfortable, those seen on the Swedish Police model below would be the answer for long rides.

Courtesy of The Sun

on 13 litres of petrol and parted company with nine Deutsche Marks. I am sorry Monsieur Pompidou, Common Market or no Common Market, I am not humoured by your price for petrol. Once you are into France the most noticeable change is the roads. Accentuated cambers, lumpy surfaces, not to mention the mini-craters which pass for pot holes, all help to keep the rider awake and the suspension busy. The BMW springs were capable of dealing with anything these roads could present. The front forks, sturdily built, are free of any head dipping tendencies, but a word of warning: Over the front mudguard is a "U" shaped chrome bridge-bar held by four nuts. Do not tighten those nuts too tightly or you will run the risk of nipping the sliders. The rear units are what all rear units should be, simple, efficient and easily adjustable. All a rider has to do is set 'em and forget 'em.

Through Strasbourg which, I am told, is a bottomless well of interesting things but unfortunately I did not have the time. (A later test run perhaps.) Following the Allemagne signs, I braked into the Customs post yet again. Green card, passport and on down the road, over the bridge. Beneath – the Rhine. It has been a big improvement since we adopted the continental type of road signs in England for we can now go abroad and at least understand the sign language if not the natives. Allowing myself to be directed by the blue autobahnen signs, I reached that concrete artery which runs from Frankfurt, south past Karlsruhe, down to Basle, front door to Switzerland.

I am not a motorway lover for a number of reasons. I was re-routed off this one and back again because of a pile-up. Many cars, I was told. Have you noticed how 'bikes never get involved in these motorway tragedies, I wonder why? However, it's undeniable that they are functional. The German autobahnen are better than our equivalents. The surface is smoother and they have uncommercial "Rastplatz" at regular intervals, where you can stop, stretch your legs, sit at the tables and benches provided, and do whatever you like; well, almost. Here the BMW was in its glory. Mile after mile never dropping below 80. It is in this environment that it brushes aside its critics. It will do a fast speed, for a long time and get you there feeling fresh. Unlike some I could mention, it does not vibrate your inside into jelly. Nor does it fall apart or blow its gaskets because it was asked to cruise at its claimed cruising speed. In fact it seems to thrive on being pushed.

If you should go down the autobahnen to Basle, please take notice of the signs telling you that the end of the motorway is nigh because this motorway ends with a little traffic-light controlled junction that would not look out of place round the back-doubles of Camden Town. If your brakes are like the BM's then you have nothing to worry about. They work extremely well but watch them on a damp morning, they can be a little severe.

My stay in Switzerland was far too short. Give me two or three weeks on a big 'bike and that's my idea of heaven. I pointed the front wheel towards home and with documents at the ready, I left this Alpine paradise. I have already gone into raptures about my run back to the Common Market capital so the answer to my acid test is 1200 miles, cruising at 80+ and, but for the starter motor, no trouble. Yes, this is my kind of girl.

Can I suggest any improvements? Well, a couple of minor things and one big belly-ache. First the little things. The handlebars are a shade too flat and narrow for me. I feel restricted, too much weight on the wrists and much too difficult to swivel the head owing to the forward inclination of the body. You must be able to move the head so you can keep your eye on the bandits at 4 and 8 o'clock. I saw a Swedish police R60 in 1967 with handlebars which must have been the last word in comfort. The other item is the horn button. For me this should be easy to reach and light to operate. The BMW's is fine if you happen to have a double-jointed thumb with well developed muscles. All too often by the time I found the button and sounded the horn, I had run over the pedestrian's left foot. But my biggest moan about the Bavarian Motor Works machines is the price. Why, why, why, does a 'bike which sells in Germany for under £700 cost £1300 in England? I defy them to satisfactorily justify that exorbitant price. I do not envy the publicity guys at Byfleet, they have simply got to be the smoothest talking pair that ever lived – "but I like them!"

A few days after I returned from my trip the 'phone rang. My wife answered. "BMW here, please can we have our 'bike back, it's wanted by *Vogue* magazine." I expect the limelight is being stolen by a fragrant smelling, beautifully dressed, comfortably built female. Like I said before, what a waste – of the 'bike I mean.

Cycle World Road Test

■ THERE ISN'T much question as to the intentions of BMW engineers when they designed their shaft drive 750. Produce a motorcycle that is capable of traveling great distances with a minimum of maintenance hassles. Allow for maximum rider comfort. Build a reputation for manufacturing one of the world's finest touring motorcycles.

The latest R75/5 is precisely that, with a touch of "superbike" thrown in. Add the bonus feature of economy to the creditable performance figures and you're faced with the prospect of riding a very desirable machine.

BeeEms don't often give way to major change, for they are a motorcycle smelling strongly of tradition. And though tradition has its strong points, the policy can often hold back progress. Fortunately for BMW, this is not the case.

BMW is apt to improve a machine during the model year, rather than wait until new model introductions. For example, our test machine, a '73 model, was the last of the 55-in. "short" wheelbase versions. Later models (unofficially 1973½) got the swinging arm extended 1½ in., a change BMW says is so minor most people will never notice.

The extra length is to allow the rider additional clearance between his shin and the Bing carburetors, long a sore spot on the 750cc horizontally-opposed Twin. An alternate bonus of the increased wheelbase should be a smoother ride at highway speeds, and BMW says that handling will not be affected. To identify models with the longer swinging arm at your local dealer, look for black fuel lines and flat black intake tubes to the carburetors. They were previously silver.

Other than those few items, new BeeEms are quite similar in all respects to last year's version. And as before, the 500cc and 600cc models are available, though the 75/5 series outsells the R60/5 by nearly eight to one. Since the difference in price between the smaller models and the larger and faster 75/5 isn't that much, most buyers are more than willing to fork over the extra cash and ride home on the top of the line. They also find that when it is time to sell, the 750 has the best resale value, though all BMWs rate high in this respect.

And once a buyer has decided that, indeed, it is a BMW that he wants, choosing the model is an easier proposition than with most brands. This is due to the fact that all BMWs are identical with the exception of bore size, carburetion, and the model nameplate on the sides of the engine.

There is no worry about which model has what, because if one has it they all do. Simply decide how fast you want to motor, the health and welfare of your checkbook balance, and pick one of seven colors.

Purists stood back and examined the new styling treatment of the 1972 models when they were first introduced. Most were unhappy. The chromed fuel tank panels had them aghast, not to mention a few other new styling "touches." But apparently the changes caught the eye of a new group of buyers, and Butler & Smith, BMW's U.S. distributor, reports that the new design was more than successful. Sales went up. The chrome stayed. The purists frowned. But they can order the "old" style 6.2-gal. tank as an option if they'd rather.

To the average rider the BMW comes off looking strange indeed. After all, *his* cylinders aren't sticking out the sides like *that*. But little do the neophytes know about the advantages of such an arrangement. The opposed Twin has those jugs sitting out there in all that cooling air rushing by. Overheat? Not on your life.

BMW R75/5

The Wayfarer's Choice

BMW R75/5

Too, routine servicing becomes a simple task, because plugs and carburetors are out in the open with nothing to prevent easy access. Even valve lash adjustment offers no problems. And BMW is thoughtful enough to include a tool kit that contains everything you'd need for a roadside repair. Among the goodies is a tire pump, extra patches, and even a rag.

Though a BMW may look like it's on the hefty side, don't let the thing trick you. The R75/5 is one of the lightest 750s available, and considerably lighter than the majority of big bore machines. This has been accomplished through the extensive use of aluminum alloys and lightweight materials such as plastic.

Again, the traditionalists will probably get upset over the mere mention of the word "plastic," but it has its good points. For example, both fenders are plastic, but you'd never know unless you gave them a tap. They are feather light, they don't dent or rust, and they'll probably never crack from vibration.

A duplex cradle type frame is used which features lightweight oval tubing in the main sections. The triangulation of bracing serving as the seat and rear fender support is removable, unlike most frames. Welding is on par with past BMWs, which is to say that it's quite good. The engine/transmission unit bolts directly to the frame with no rubber mounts or cushioning whatsoever.

The 745cc flat Twin is mechanically interesting to look at. In fact passers-by with no interest whatsoever in motorcycles are likely to pause and ponder a moment.

Models from a few years back had lots of iron bits and pieces throughout, but the modern day BMW has made the switch to alloys. Now we find aluminum cylinders with a cast-in steel liner for better longevity. The main engine casting is alloy, as well. Though the unit appears as though it is on the tall side, the upper portion is comprised of an air chamber and housing for the electric starter motor.

The heavy pieces such as the forged steel crankshaft, steel connecting rods, and camshaft, are positioned down low. So in spite of the way it looks, the center of gravity is where it should be. And to help with cornering clearance, BMW raised the cylinders slightly a few years back, and that's where they've been ever since. The change made it necessary to place the camshaft below the crank, rather than above, but now the cam benefits from better lubrication.

Bend down and look at the bottomside of the engine cover, and there is a large cast aluminum sump cover with cooling fins. This aids in dissipating heat from the nearly five pints of oil contained in the oiling system, which is circulated by a Eaton-type pump through a full-flow oil filter. Since plain-type bearings are used in the engine, an efficient oiling system is imperative.

A double-row chain connects the crankshaft and cam together, and features a tensioner unit to help eliminate any vibrations which may be present. Doing away with chain slack also helps somewhat in making the engine mechanically quieter, and quiet is another one of those BeeEm traditions.

This is one motorcycle that was silent before it became fashion, and now when other manufacturers are working feverishly to meet new requirements, BMW finds itself concentrating on other items. They have a right to look smug.

It's interesting to note here that the 750 BMW has always been a strong performer, but has always been left out of the "superbike" classification. Now suddenly, many of yesterday's "super" machines are finding themselves with far more restrictive muffling and air intake tracts, due to the new laws. As a result, their drag strip times are falling off drastically, and again, BMW looks pretty smug. If anything, they're going faster.

An automotive-type single plate dry clutch is continued on the latest models. Lever pull is light and release is positive. So positive, in fact, that power shifts produce impressive wheelies from first to second at the dragstrip. And once again, BeeEm has reduced the size of its flywheels, so the old BMW "clunk" during shifts isn't as prevalent. With the lighter flyweels, engine rpms slow much more quickly and as a result, engine and transmission speeds synchronize just that much faster. Transmission gears were beefed up last year and no changes have been made since.

Naturally, the feature that makes the BMW stand out in the local crowd of wayfarers is shaft drive. There is no chain to lubricate, adjust, or sling oil on rider and machine. This makes for an ideal situation whether you are on a long trip or a short one. The driveshaft nestles inside the right hand swinging arm member, and connects to a hypoid gear at the rear wheel. The gear rides in an oil bath and is sealed, so the rider can virtually forget about it.

Rocking the 750 off its centerstand and climbing on feels funny if you've never perched yourself on a BMW. The cylinders feel as though they'll get in the way and they give you an impression that the machine is huge. But that's only if you're staring down at them. And since most riders don't stare down when they're in motion (we hope), they forget about the fact that they're there.

Since there is no key switch on a BMW, the ignition is turned on with a little plug-in handle that fits in a hole on the headlight nacelle. In the "down" position the ignition is on, and turning the handle left or right operates parking or running lights. This is one feature on the BMW that is annoying, and it makes theft easy since anyone with one of those plugs can start your motorcycle and ride away. A key switch is a much better way to go.

Some choke has to be fed to the Bing vaccuum-type carburetors if the engine is cold, and depressing the thumb operated button has the powerful electric starter motor churning several revolutions before the Twin fires. Once warm, however, the engine starts with much less reluctance.

At idle the engine is quiet; blip the throttle and it stays quiet. But there is that peculiar torque reaction in the lateral plane when rpm builds. It's very strange for someone used to a more conventional machine.

Step on the shifter pedal with your left foot and pull away; there is a shake throughout the bike at low rpm, but that disappears as revs build. And it's not the kind of shake that gets to the rider, except for the fact that images do not focus in the rear view mirrors. Only four speeds with this one, and that's all the BeeEm needs to do business.

Around town you can stay in first gear and not bother anyone because of noise, but you can also lug down the 750 in a higher gear if you don't mind the shakes. All controls are properly spaced, but the Magura levers require some reaching with the fingers to grab hold of them firmly.

Front fork action is strange. Apply the front brake and there is 8 in. of soft compression. Yet, on certain street surfaces, the forks don't seem to budge an inch, and the corresponding ride is harsh. The forks take the big bumps in stride, but ignore the little ones completely.

At the rear, shock action is close to perfect over any surface. They even place a handle at the bottom of each unit so the rider may change the settings without hunting for the proper tool.

On some lengthly trips, staffers began to appreciate what BMW smoothness means. Long rides and fast travel magnifies the comfort. And related handling and braking is up to the overall performance of the motorcycle. There are minor flaws and annoyances that keep the R75/5 from perfection, but overall refinement approaches excellence.

But don't just pass judgment on a 5-mile tour around the block. Motor along for awhile. The bugs may go "splat" and the trucks may go "whoosh," but the R75/5 doesn't give a damn. The BMW R75/5 is one sweet ride to anywhere.

BMW R75/5

SPECIFICATIONS

List price	$2190
Suspension, front	telescopic fork
Suspension, rear	swinging arm
Tire, front	3.25-19
Tire, rear	4.00-18
Brake, front, diameter x width, in.	7.9 x 1.2
Brake, rear, diameter x width, in.	7.9 x 1.2
Total brake swept area, sq. in.	59.57
Brake loading, lb./sq. in. (160-lb. rider)	10.02
Engine, type	ohv opposed Twin
Bore x stroke, in., mm	3.22 x 2.77, 82 x 70.6
Piston displacement, cu. in., cc	45.2, 745
Compression ratio	9.0:1
Claimed bhp @ rpm	57 @ 6400
Claimed torque @ rpm, lb.-ft.	43.4 @ 5000
Carburetion	(2) 32mm Bing constant velocity
Ignition	battery and coil
Oil system	gear pump, wet sump
Oil capacity, pt.	4.75
Fuel capacity, U.S. gal.	5.0
Recommended fuel	premium
Starting system	electric, kick
Lighting system	12V alternator
Air filtration	treated paper
Clutch	single-plate, wet
Primary drive	none
Final drive	shaft and bevel gear
Gear ratios, overall:1	
4th	4.80
3rd	6.00
2nd	8.25
1st	12.47
Wheelbase, in.	56.5
Seat height, in.	33.0
Seat width, in.	13.0
Handlebar width, in.	28.5
Footpeg height, in.	13.5
Ground clearance, in.	8.5
Curb weight (w/half-tank fuel), lb.	437
Weight bias, front/rear, percent	44/56
Test weight (fuel and rider), lb.	597
Mileage at completion of test	1468

TEST CONDITIONS

Air temperature, degrees F	80
Humidity, percent	54
Barometric pressure, in. hg.	30.00
Altitude above mean sea level, ft.	350
Wind velocity, mph	6
Strip alignment, relative wind:	

PERFORMANCE

Top speed (actual @ 7000 rpm), mph	109
Computed top speed in gears (@ 7000 rpm), mph:	
4th	109
3rd	88
2nd	64
1st	42
Mph/1000 rpm, top gear	15.5
Engine revolutions/mile, top gear	3846
Piston speed (@ 7000 rpm), ft./min.	3232
Lb./hp (160-lb. rider)	10.47
Fuel consumption, mpg	44
Speedometer error:	
50 mph indicated, actually	45
60 mph indicated, actually	56
70 mph indicated, actually	66
Braking distance:	
from 30 mph, ft.	36
from 60 mph, ft.	143
Acceleration, zero to:	
30 mph, sec.	3.9
40 mph, sec.	4.6
50 mph, sec.	5.7
60 mph, sec.	6.2
70 mph, sec.	7.5
80 mph, sec.	8.5
90 mph, sec.	10.9
100 mph, sec.	15.0
Standing one-eighth mile, sec.	9.97
terminal speed, mph	87.46
Standing one-quarter mile, sec.	13.55
terminal speed, mph	97.93

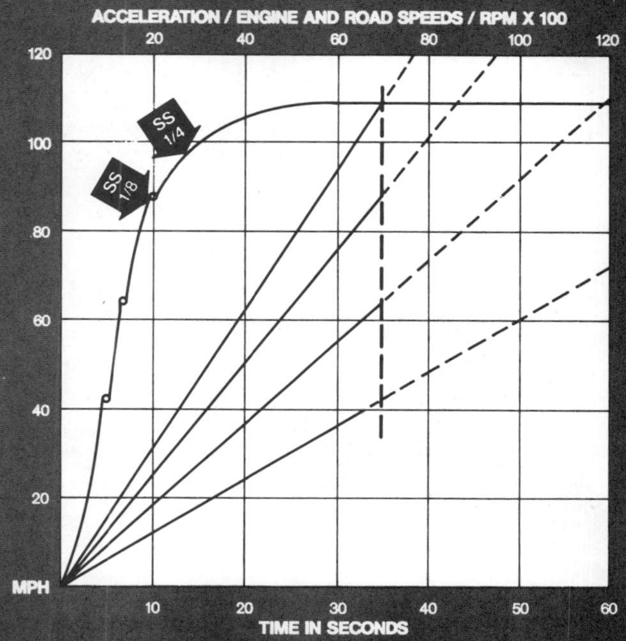

CAMPING WITH a motorcycle? Why not? The excitement of twisting through a winding mountain road with unobstructed vision, then relaxing beside a campfire at the end of the day just can't be beat. With proper planning and a wise investment in equipment, camping by motorcycle can provide many memorable weekends and vacations.

My wife and I decided that a trip from Dallas, Texas, to the Pacific Northwest on our new R75/5 BMWs would be a natural for our vacation. Our inventory of camping equipment was not very good. So, we made a list of items that were needed. What a list! How would we ever get all that on two wheels?

The most exciting part of preparation turned out to be the selection of equipment. We found that firms which specialize in lightweight backpack equipment have a lot to offer motorcyclists. Backpacking and motorcycle camping have one thing in common—the challenge of getting items of necessity into a relatively small space and yet restricting weight to the least amount possible.

Remember: you get what you pay for. The initial investment will pay off with equipment that will give years of durable service and pleasure. We found that our list covered four basic areas of necessity: camping equipment, clothing, food, and tools and spare parts.

Camping equipment—As far as tents go, the three-man Eureka drawtight tent came closest to our needs and yet did not stretch our checkbook too far out of shape. The decision was mainly economic but size also entered into it. The Eureka folded down to a little larger and heavier bundle than I wanted.

(The other choice was a Gerry Fortnight nylon backpack tent for almost twice the money). The Eureka, although it is called a three-man, will sleep four adults providing all are under six feet tall. This size tent was important to us because we wanted to take another couple with us on future weekend trips.

We chose down-filled sleeping bags mainly because they compacted into small rolls, thereby reducing the bulk that a dacron-filled bag takes. A goose down bag also offers the best insulation by weight.

Getting the bags stuffed into the

CAMPING

Proper Equipment Plus A BeeEm Equals A Sweet Ride To Anywhere / By David Peterson

storage bags proved to be a trying experience. I worked up a good sweat on my first few tries. If I had it to do over again I'd buy stuff bags that lace up into an even smaller roll.

Other equipment included a hatchet, candle-lantern, and ground insulation pads (used under the sleeping bags for insulation and cushion). Lanterns with mantels won't work too well. A friend took his Coleman out on a short weekend trip only to arrive at his destination after dark with a broken mantel and no spare in sight. Setting up camp was no fun in darkness.

We also took along a flashlight, AAA triptick, maps, camping guides and tour books. Toiletry articles included a very useful item—Wash 'n Dry folded moist towels sealed in a very thin container. These can be used once and discarded. I even found room for my 35mm camera.

Food—After a hard day's ride even freeze-dried meat and vegetables tasted great. Granted its quite a switch of menu from the food we are used to at home.

Food preparation can present difficulties—especially cooking with a one burner Bleuet stove when gusts of wind blow the heat and flame in every direction but up. With some patience and an ingenius windscreen we overcame what otherwise could have been a disaster.

We found it best to prepare supper at the campsite. Breakfast and lunch were best eaten out of restaurants rather than a cardboard box. This provided us with more time during the prime of the day for traveling. And besides, if we had prepared breakfast at the campsite, it would take extra time to clean up the mess. Food taken with us consisted of high energy foods, tropical chocolate, fruit bars, beef jerky, freeze-dried meals, salt and pepper, cooking oil and coffee.

Cooking utensils consisted of the Bleuet one burner stove and a camping set of nesting utensils that included pots, pans, plates and cups. A set with a Teflon fry pan overcomes the problem of food sticking to the pan.

A few freeze-dried food items are normally found in local supermarkets. However, if your taste does not adapt to a bland diet, you'll need to find a source that caters to backpackers. Recreational Equipment, Inc., 1525 11th Ave., Seattle, WA 98122, has by far the best selection and stock on hand. Occasionally sporting goods stores and other mail order firms carry several lines such as Wilson Freeze Dried Foods, Seidel's, Trail Packets and Rich Moor lightweight foods. Its best to buy and sample the meals and foods that appeal to you before stocking up on too much. An alternative to loading down with freeze-dried food would be to buy fresh food for each meal you are going to prepare.

Clothing—Bike camping is very susceptible to extremes in weather. Since Mother Nature does not always cooperate, clothing becomes an important factor. We went prepared with rainsuits, thermal underwear, changes of regular garments, mittens and leather jackets. It all paid off. Our second day out was spent driving for hours in the rain. The winding mountain roads of Colorado in combination with wind, rain and mud was really a challenge. Visibility created a real problem. I wished I had windshield wipers on my goldrim glasses to keep the mud and rain off of them.

The following day was beautiful, but we drove into what seemed to be an eternal deep freeze—the 12,000 foot elevation of Independence Pass near Aspen, Colo. We were thankful for the thermal underwear, sweaters, leather jackets and ski mittens.

Dehydration is a critical problem in hot weather. Clothing, especially leathers, oddly enough help retain some body moisture. But the combination of direct sun and dark colored leathers may become unbearable in the heat of the day.

Entering the Columbia River basin was like going from the freezer to the frying pan. The prolonged summer daylight hours and 105 degree heat made the salt tablets and a canteen of water to help prevent heat exhaustion a necessity.

Surviving the extremes is the key. It takes preparation and conditioning. Our suggestion to anyone making a cross-country trip, as far as the weather is concerned, is to be prepared for anything!

Tools and spare parts—Every cycle has its problems and it is best to travel prepared. A friend of mine set out for Nashville, Tenn., from Dallas and his chain broke. He had a master link that he faithfully carried but lo and behold, he didn't have a chain breaker to remove the broken link. Fortunately, he was able to muster help from a nearby Honda dealer and was on his way a little worse for wear but a bit wiser. Ever since then, when setting out for a relatively long trip, the first thing in his saddlebag is a spare chain.

The BMW's driveshaft configuration eliminates the chain problem. However, we went prepared with a minimum of spare parts—plugs, points and a spare tube, and a can of self-vulcanizing tire seal used to inflate a flat tire and seal punctures that have not seriously damaged the tube. Tools included the tool kit, tire patches and a pump—all of which came with the bike.

I also threw in a continuity light that could be used as a timing light. This was a homemade affair-two pieces of lamp cord, each about 2-ft. long, attached to alligator clips on one end. The other ends were soldered to a small 12V

Continued on page 141

FIFTY YEARS

● STILL SMOOTH
● STILL RELAXING
● STILL SUPERB

While other manufacturers chase themselves round in circles searching for the next "bigger and better" gimmick for the next generation of super duper bikes, BMW have been happy to stay with their well-proven range.

OK, so they did a little detail redesign and restyling a short while ago, and as far as price goes BMW were in the super bike bracket long before the phrase was even invented. But for sheer performance their machines just aren't as rapid as most others and they have three models which are very, very similar to one another.

The 500 cc subject of this test is around 12 lb. lighter than the 750 we tested last year and some 10 to 15 mph slower. Most of the time you did not need to know whether it was the 500, 600 or 750 you were riding.

Obviously BMW aren't counting on performance to sell their machines, their strong point has to be reliability, longevity and comfort—"driveability", if you like. I'd go along with that and add that any of the BMW range make the best long-distance tourers going. It still doesn't explain why they make three models which are so similar to one another, unless

on! BMW R50/5 TEST

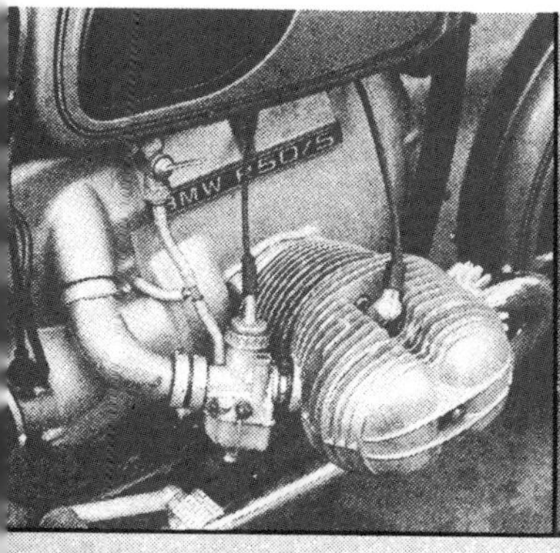

BMW's flat twin motor—a smooth flexible lump that doesn't like to be hurried too much

It doesn't look too impressive but the brake is powerful and matches the handling nicely

The controls are just right, and easy to operate, but the instruments could be better placed

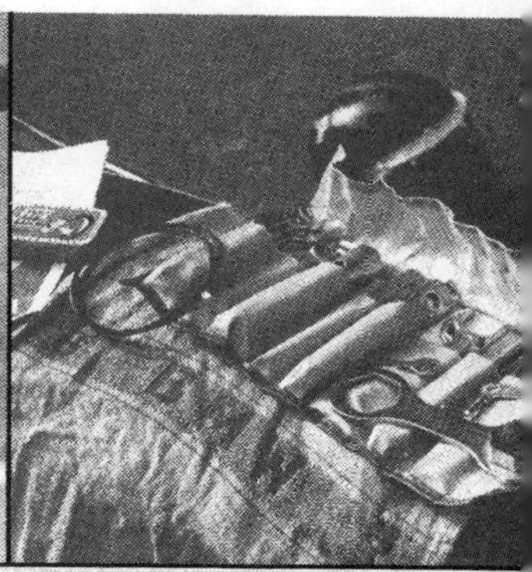

This excellent toolkit i typical of the thorough attentio to detail around the machin

on the German home market there is a greater price differential. If the 500 could be considered an economy version, the set-up would make more sense. Well, it does have a lower compression ratio . . .

We found that riding the R50/5 under everyday commuting conditions didn't tell us a lot about the machine. It compares reasonably well with other models of the same size; on the credit side it is docile and quiet, on the debit side the gearshift made louder clunks than any Bee-Em I've ridden yet.

It wasn't until the machine was headed out for a long run that it began to show its real value. On one occasion a minor crisis forced me to ride a hundred miles or so one dark, cold, very wet night, without adequate waterproof clothing. I was thankful we had the BMW; the journey still wasn't exactly pleasant, but on most other machines it would have been downright painful.

Even as I write this, Chas Deane has just got back from the Island, having taken part in BMW's attempt at the Maudes Trophy. The weather conditions there were pretty grim, but his woe-begotten tales confirm all I was saying about the BMW's driveability:

First, the machine is comfortable—the riding position is just right and the seat supports you in a relaxing way. Second, the handling is very good. On wet or dry roads the BMW will cope easily with all kinds of surface, and you can put it just where you want it. Ridges, grooves or white lines which upset a lot of other machines, leave the BMW ploughing steadily along.

The longer you ride the machine, the more you realise how effortless it all is—what BMW give away in terms of performance they more than make up for in their easy handling.

On the 500 acceleration and top speed are below average, under fairly neutral conditions it will pull a maximum of just over 90 mph. If you can coax it up to peak revs in top, it is geared to reach something approaching 100 mph. This isn't particularly astounding, but it will hold a steady 80 mph or so no matter what—and it gives the impression of being able to keep this speed up all day.

We found that the fuel consumption varied little from motorway cruising to town riding, ranging from the mid-fifties to a shade over 60 mpg.

Before you even get down to details, this adds up to a fine tourer. When you do start looking at the details, the picture gets better and better.

In my opinion the machine is one of the finest looking motorcycles ever built and the standard of finish matches its rugged good looks. Details around the machine have been carefully thought out and designed for the rider who *uses* his motorcycle. The comprehensive, high quality toolkit, for example sits in a tray under the seat which can be locked down. A cloth and a puncture repair outfit are provided and there is still enough room to carry a spare set of light bulbs, plugs, points, etc.

On the subject of carrying things, I was disappointed that no carrier or pannier is fitted as original equipment. If this is a machine intended above all else for long distance riding, then it needs carriers more than most others. When one has laid out all that cash one shouldn't have the extra outlay and trouble involved in fitting carriers. Also, now that the appalling crash helmet law has become fact, machines need stowing space more than ever.

Another detail which is basically a good idea is the fly-up propstand. This is spring-loaded, so that when the weight is taken off the stand it automatically folds away. In practice it didn't work too well, being unreliable if the machine was on any kind of a slope and tending to lift if one leaned against the machine.

All of the controls and switches were well placed and easy to work, even in bad riding conditions, and the electrics are powerful enough to be compatible with the machine's performance. A headlamp flasher, which I regard as more of a necessity than a luxury, is fitted, but there is no ignition kill switch. The composite tacho/speedo spoils the arrangement up front, it doesn't look too good and it isn't particularly easy to read. And they still fit that idiotic ignition switch.

The electric starter is quite powerful, and I never had any problems starting the BMW. No choke is fitted, and from cold both carbs have to be tickled and the engine spun for three or four seconds on a tiny amount of throttle. If the motor didn't pick up, I found it better to give the starter several short bursts rather than keeping it spinning. Usually on the second or third go, one cylinder would fire up and by gently touching the throttle the other would chip in. Then it needed about thirty seconds running around the 2000 rpm mark before it would give good throttle response.

Once under way the motor is smooth and flexible, with plenty

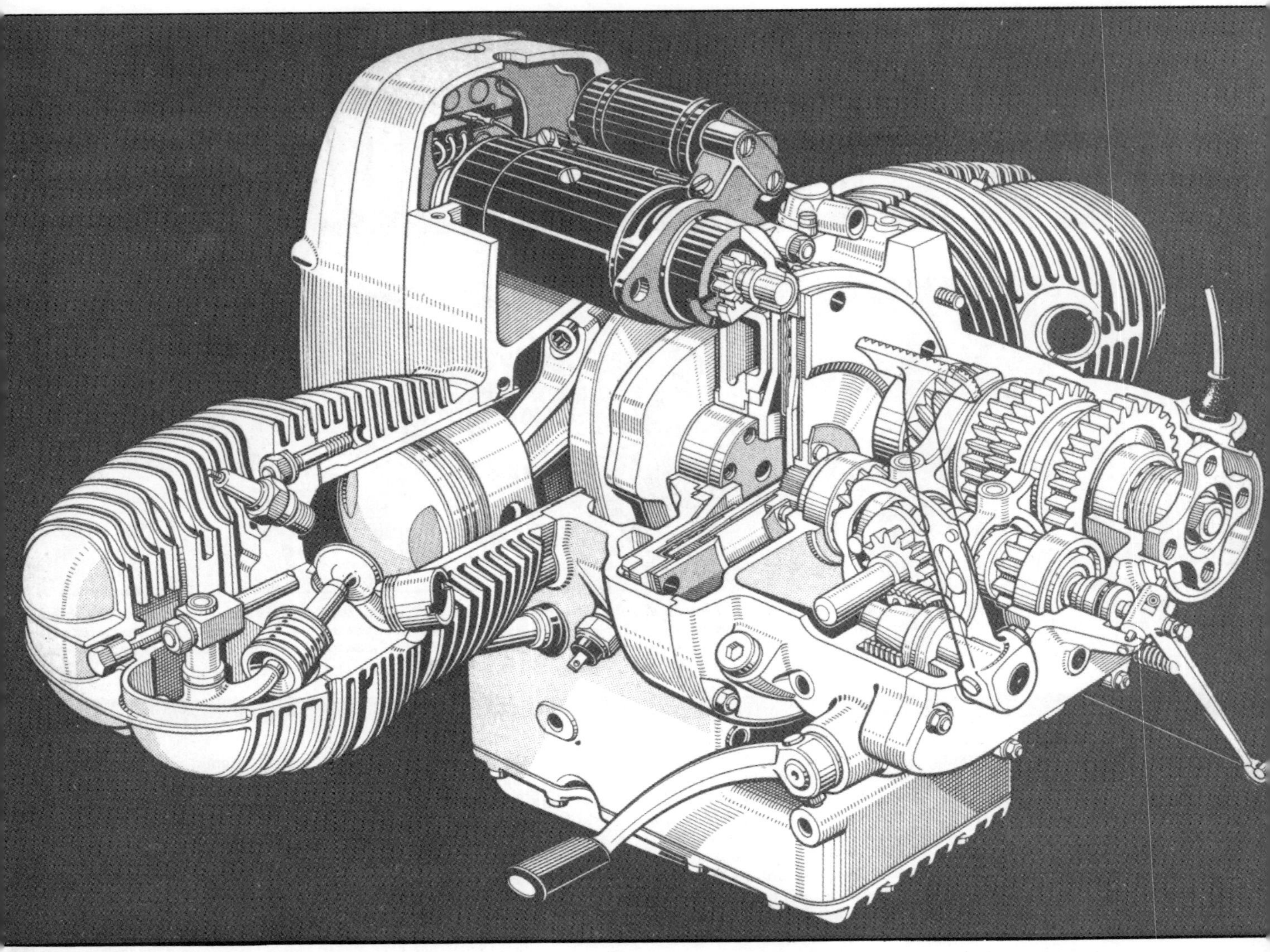

BMW TEST

of low down torque and no noticeable "cam" effect. It gives that confident, long-legged feel which tempts you to hang on to top gear and rely on the engine's smooth pulling power.

This is fortunate, because the BMW gearbox doesn't encourage enthusiastic cog-swapping. All of the BMWs I have ridden have had clunky gearchanges, although with a little concentration a silent shift is possible. This 500, however, seemed to have a mind of its own and clunked no matter what. Then, just when I'd given up trying to make silent changes, it would occasionally slip into the next gear with barely a click, as if to remind me that it *could* be done.

As with other BMWs we've ridden, this one developed a clappity tappet—it would usually start clattering after a long fast run, and then it was only intermittent. That was the sole fault which showed up during the test—there was no vibration, no oil leaks, nothing to get upset about at all.

The roadholding is excellent, and I've already gone on at great length about the nice handling. The BMW doesn't like being thrown about, but it will go where you point it and stay on line without any nervous twitching or shaking of the head. Ground clearance is very good. The stands will graze the floor, but you ought to start looking for a soft spot in the hedgerow before you need worry about anything else touching. Unfortunately with all that engine hanging over the side, the dire consequences of a terminal slide are always in the back of one's mind and this, along with the general character of the machine, damps down any tendency to scratch around corners.

The Metzeler tyres have been criticised by other testers, but I have no complaints at all. I like the feel to the handling which is given by fat tyres and on wet roads I felt more confident on the BMW than on any other roadster I've ridden.

Braking on the BMW presented no problems. The twin-leading-shoe job at the front goes about its work with no fuss and even a lot of hard work didn't prompt it to fade. When the back brake was adjusted up to suit my foot position it was a little too keen. Backing it off some took the edge off its sharpness, but made it that much more uncomfortable to operate.

Two-up riding doesn't give the 500 any trouble and it is comfortable enough for the passenger even over long distances. The pillion footrest gives the passenger a comfortable position, but it is too close to the main footrest, and the passenger has to ride with the ball of the foot on the rest. It would make things a lot better if the footrest were moved back a couple of inches. On the subject of footrests, a word of warning—don't wear flared trousers outside your boots! The reason is that as the machine comes to rest and you're about to put a foot down, the said trouser neatly hooks itself over the tickler on the carburettor; I leave you to imagine the rest.

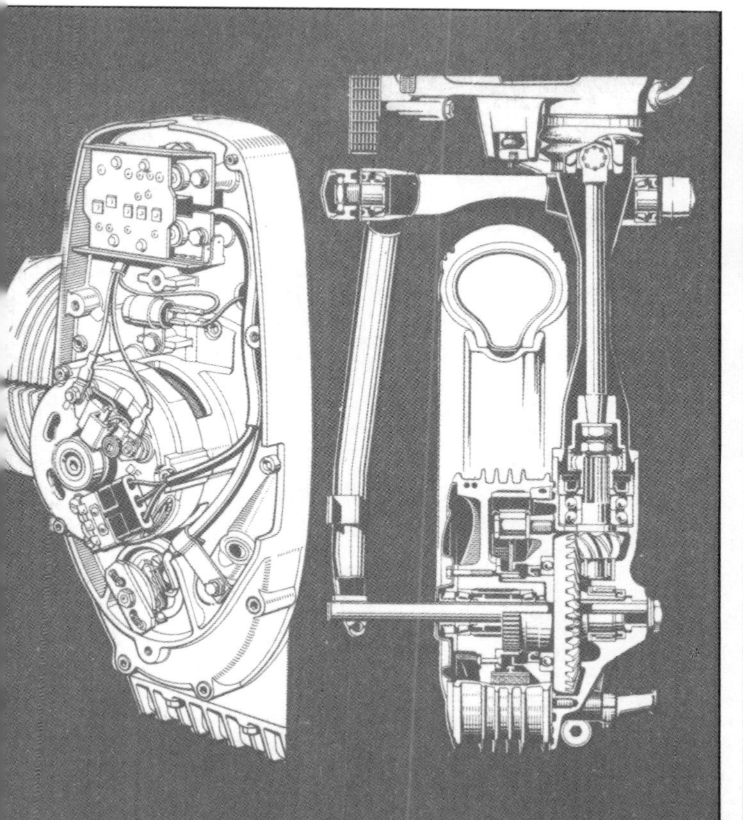

Cutaway drawings show the workings of BMW's now classic engine and transmission layout. Far left: The engine and gearbox, showing the single plate diaphragm clutch, with the starter motor sitting on top of the crankcase. The gearbox is the crossover type. Above left: At the front of the motor is the three-phase Bosch alternator, rectifying and regulating system and the camshaft-driven contact-breaker. Above right: The gearbox output shaft couples through a UJ on to a shaft running inside the swinging arm. The final reduction is made through a spiral bevel crown wheel and pinion.

The BMW, although it is the "baby" of the family, is only really at home on long main roads. Then you can feel the true concept of the machine, whistling quietly and smoothly along and shrinking the miles as easily as machines with far greater performance. Just as the engine is understressed and never seems to be working hard, so the chassis is equally smooth, giving a beautiful compromise between comfort and predictable roadholding. That, despite the GT labels on other machinery is what *gran turismo* is all about.

There have been few machines which have been able to compare with BMW's smooth, silent power and comfortable, easy handling. The only real point against the machines are their high cost and the cost of parts—but there seem to be plenty of people who are prepared to pay for their motorcycling.

Then, again, the BMW concept could be considered old-fashioned: the layout of the BMW has been with us for a long time now, the 1923 side-valve R32 500 bears very strong resemblance to today's machines, and BMW have always kept the flat-twin, shaft-drive concept.

Where do they go from here, though? Fifty years is a lot of detail redesign, and there can't be a lot left for them to do to their twins. The layout doesn't lend itself to the addition of extra cylinders, and as BMW don't seem interested in all-out performance, a bigger or more powerful motor wouldn't be achieving anything.

So they appear to be left with a range of three very similar machines. One alternative does spring to mind—the chassis and transmission of the BMW seems eminently suitable for a rotary engine. A two or three rotor unit could be installed, as without a space-consuming crankshaft there would also be room for a liquid coolant system. Maybe we will see some changes from BMW . . .

JOHN ROBINSON

ROAD TEST BMW R50/5

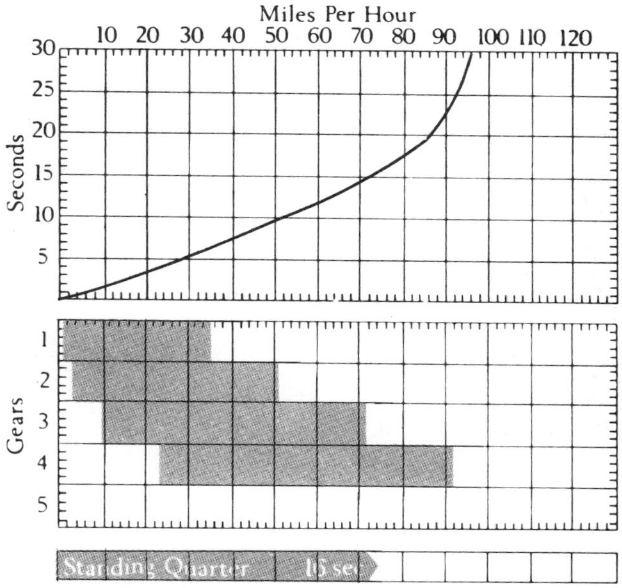

Fuel consumption (average) **57 mpg**
Braking from 30 mph **29 ft**

SPECIFICATION

Engine: Horizontally opposed OHV twin. Bore × stroke 67 × 70.6 mm, capacity 498 cc. Compression ratio 8.6:1. Carburettors, twin 26 mm Bing concentric float type. High pressure oil feed to bearing surfaces from Eaton pump. Claimed output 36 bhp at 6600 rpm.
Transmission: Single plate clutch with diaphragm spring, four-speed crossover gearbox. Final drive by shaft and spiral bevel gear, overall ratios: 13.84, 9.18, 6.68 and 5.35:1.
Electrics: Bosch 12 volt ac/dc lighting, supplied from crankshaft-mounted alternator, feeding battery via rectifier and voltage regulator. Ignition by twin coil and contact breaker. Indicators and starter motor fitted as OE.
Wheels: Welled rims, 1.85B × 19 front, 1.85B × 18 rear. Metzeler tyres, 3.25 × 19 front and 4.00 × 18 rear. 7.9 in. brakes, with 2ls front.
Suspension: Front, telescopic forks, rear, swinging arm with adjustable damper units.
Dimensions: Seat height 33.5 in., wheelbase 54.5 in., ground clearance (with 165 lb. payload) 6.5 in., kerb weight 408 lb.
Price: £899·98.
Importers: BMW, Great West Road, Brentford, Middlesex.

BMW R75/5

For most enthusiasts no consideration of touring bikes would seem complete without inclusion of the BMW. Over the years the BMW has developed its reputation as a good touring machine. Consequently, the opposed cylinder twin has come to mean just one thing—smooth comfortable, long distance riding. It has been, and continues to be a specialist.

It would obviously be foolish for BMW to try and appeal to the performance minded public. This twin never has been and never will be (without modifications which could only be purchased with the loss of some of its mystique) a performance machine. Wisely, we think, it hasn't attempted to seek entry into the somewhat nebulously defined (for the back markers anyway) super-bike performance corral. It lacks the intangible ingredients, even though its quarter mile performance matches those of three Japanese bikes (Honda, Suzuki and Yamaha) that many would readily include in any super-bike grouping.

With the recent entry of the Japanese manufacturers into the big bore field, the touring market has been expanded. Almost any 750 can obviously be used for touring with varying results. (The only popular 750 we excluded from our test was the Kawasaki Mach IV. This is one super-bike with all the emphasis on super alone rather than the duality we feel is necessary to double as a touring mount.)

In effect the BMW (and Harley-Davidson Electra Glide) are being challenged by machines that can also quench the thirst of riders with split personalities, those who like touring as well as gassing it. This widening of the market selection for the rider who primarily is interested in touring is one of the reasons for this comparison test. With the BMW we were able to take what many consider to be the Mercedes Benz of touring (this would probably lead one to picture the Electra Glide as the Cadillac of touring) and see how it fares with the new competition from across the Pacific.

In appearance the BMW is a trip in its own, with the two cylinders protruding out both sides and the bulbous looking aluminum cover above the engine that houses an air cleaner, starter

and electrical system. Contrary to its visual impression the R75/5 weighed in with 4½ gallons of gasoline at 465 pounds. *It was the lightest of all test bikes. The Suzuki was the heaviest at 549 pounds. Next to the BMW was the Triumph at 489 pounds.*

After glancing a critical eye over the various components such as castings and the quality and finish of the paint and chrome we must conclude that BMW has done a fine job. The interesting thing to note here however, is that this quality of finish no longer stands out head and shoulders above the competition. Bavaria has slipped, and the rest of the field is catching up.

Appearance and finish aren't the only considerations when looking for (or testing) a touring type motorcycle. In the past we have logged several 5000 mile test rides on various machines, but none of these long range evaluations included a BMW. However, with all the miles we have put on other test units, and all the miles we have driven from one coast to the other in four-wheeled vehicles, hauling competition two wheelers, we have yet to see the trusty BMW parked by the road side in need of mechanical assistance. We have seen many other types of machines in this predicament. Part of this can be explained statistically because BMW is by no means a world sales leader, but still one would think that we would have seen at least one in the last six years if there were any to see. This just reinforces our belief that the BMW reputation for reliability is justified.

The R75/5 comes equipped with two types of starting systems, the old reliable kick starter and the easy, convenient electric starter. *Only the Triumph was without electric starting.* On cold mornings it is necessary to leave the throttle closed, the choke on and only then tap the button. Within a few seconds the engine comes to life. On warmer days, once the engine has been started, the choke is not necessary. However, a little throttle does help.

The kick starter method is something else again. Luckily, if it is necessary to use this method, the engine starts quite easily. The kick lever seems awkwardly conceived. To operate the kick starter it is necessary to kick to the side of the machine rather than toward the rear. However, the kick lever does tuck in out of the way and doesn't rub against the ankle.

Once started the R75/5 requires several minutes of warm up time before the engine will take the throttle without gasping. It will idle with the choke on, with no outside assistance from the rider.

The BMW's powerplant is super quiet, *second only to that of the Suzuki.* As far as overall noise reaching a db(A) meter the BMW is decidedly the quietest. The mechanical noise from the BMW seems to emanate

from the valve train. Once moving, however, this noise, and in fact the exhaust noise also, is impossible to distinguish above that of the wind.

One acoustical problem, as well as operational irritant, is what occurs when shifting gears. There is an extremely noticeable clunk that is not only easily heard but also felt as the whole bike seems to shudder for an instant. It occurs on each and every shift, regardless of the gear being selected. First time riders, accustomed to the smooth transmissions found on almost all Japanese bikes, will certainly require a period of adjustment. As bad as it is, the clunk this year seems less when compared to units of pre-

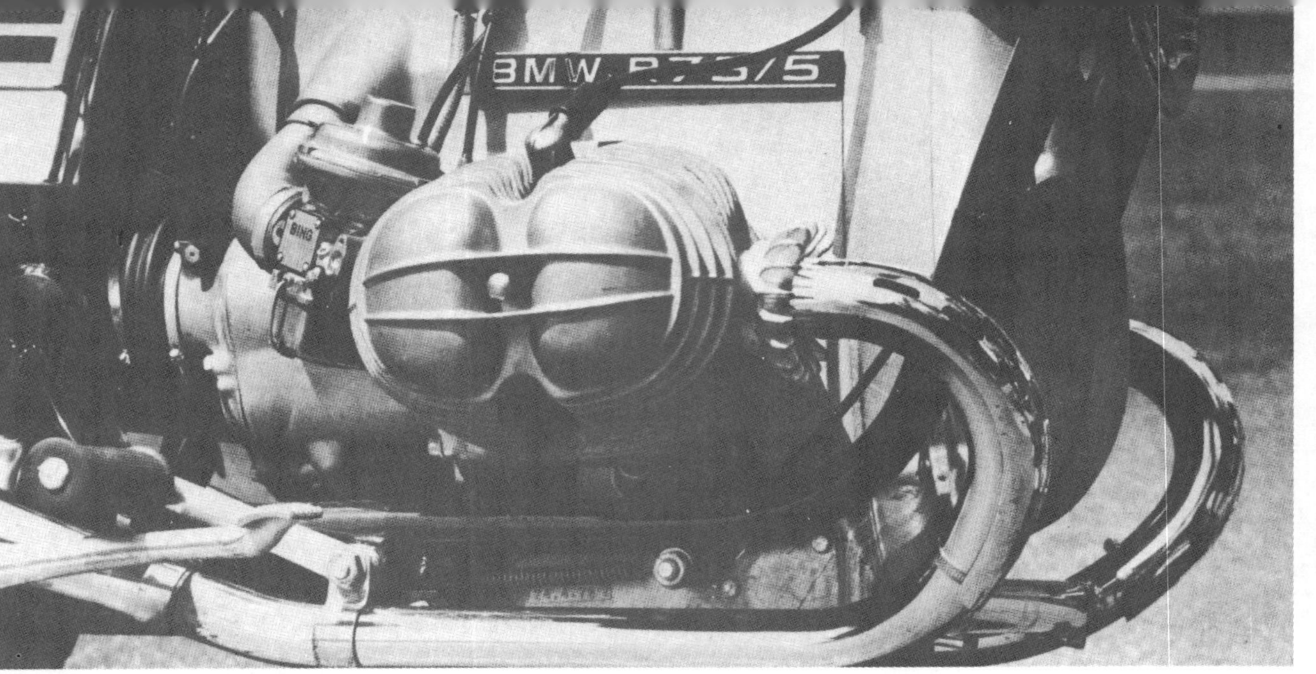

vious years. The clunk can be attributed to a design deficiency, or perhaps we should label it an exercise in engineering over kill. However, the chance of gearbox failure is almost nil, because of the design. Each gear is extra heavy duty, wide, and obviously cut so that shifting into the next higher or lower gear is not easily facilitated. This produces the clunk. Once the gear is engaged though we cannot envision the gearbox slipping out of gear. Because of this design it is necessary to shift slowly and deliberately. It is necessary to use the clutch. *The Triumph and the Yamaha were the easiest and smoothest to shift, followed closely by the Honda.*

The clutch was another item that we felt left something to be desired. *As it did with the gearbox, the Triumph also proved to have the best clutch action.* The clutch on the BMW was rather harsh and particularly quick. That is, it was either engaged or disengaged. Because of the BMW's power range it wasn't nécessary to slip or feather the clutch. Once the engine was above an idle it would pull up to its maximum rpm at 6400. *We felt that the torquiest bike was the Yamaha. Closely following were the Z1 and also the BMW.*

High speed roadability was another category that we took into consideration when evaluating the machines. Up to 80 mph we felt that the BMW was as stable as anything we tested. However, once that figure was exceeded our test unit tended to move around slightly. This was unnerving rather than dangerous. *Traveling at over 80 mph in a straight line found the Kawasaki Z1 on top followed by the Honda, Yamaha and Suzuki. Next in order was the BMW followed by the Triumph.*

When accelerating hard, particularly coming out a corner on several occasions, we felt the rear end start to slide momentarily. We feel this is due more to the design of the tire than anything else. We can suppose that since the BMW is built primarily as a straight line touring machine, it is not meant to go slithering fast around corners. *Tire traction on the other units proved to be excellent.*

We did notice an adverse reaction to freeway rain grooves on the BMW. As we have tried to point out in the past, we feel that the major cause of the front end oscillation encountered in this situation is due to not only the selection of tire tread design but also the relative size of the front and rear tires. During high speed cornering we noticed a slight twitch from the R75 which we felt was caused by the soft suspension on the unit. *Perhaps the two bikes most insensitive to high speed cornering were the Honda and the Kawasaki. Both of these units went around the corner rock steady.*

Slow speed cornering on the other hand will probably require a getting used to period for the average rider. If the throttle is opened quickly at a speed of 35 mph or less the torque reaction gives the bike a feeling of being lifted. This same lifting sensation is noted when shifting because of the unloading of the front suspension by the torque reaction. Actually, this sounds a lot worse than it actually is. Once the speed gets above approximately 35 mph this trait is no longer noticeable.

Another area requiring investigation is the ability of the brakes to perform well under general conditions, the everyday on the road stops that one may encounter. Although the BMW was the only bike without a disc brake fitted up front, it *and the Suzuki with dual disc brakes* were our choice for the best all around braking performers. *Surprisingly enough, because of previous impressive performance, the Honda was the most difficult to stop.*

Even though the BMW was fitted with a drum brake front and rear we were quite surprised at the results of the panic braking test. *As might be expected the Suzuki captured top honors in both the 35-0 and 65-0 tests.* We knew that the BMW was the lightest of the bikes, but we also realized that it had theoretically the least efficient brakes. So much for theory. The BMW was the second quickest stopper from 35 mph with a mark of 39.6 feet. The R75/5 was the fourth quickest stopper from 65 mph, *just two feet behind Yamaha at 143'5". As far as stability while braking we feel that the Kawasaki was tops. The BMW rates alongside the Honda, behind the Z1.*

When getting on the brakes hard the front forks tended to compress nearly the full length of their travel. This gave us some insight as to the quality and comfort of the ride that could be enjoyed on the twin.

There is no way of getting around the fact that of the six bikes we tested the BMW had the best suspension as far as touring type comfort is concerned. *The most uncomfortable we felt was the Kawasaki Z1 because its suspension units, particularly in the rear, were quite stiff.* The BMW has relatively soft units, with a considerable amount of travel and good damping. The front forks are allowed to move 8.2 inches while the rear shocks move 4.9 inches. These travel figures are greater than those of any other machine we tested.

Aside from the very slight front end squat and the slight wiggle when rounding high speed corners this soft suspension with long travel was a definite advantage. Only once or twice did we ever get bounced off the seat due to hitting a large bump. Generally speaking the front tended to smooth out most of the road surface irregularities and the rear suspension did, likewise.

Once we got out on the highway it was quite evident to us why the BMW has enjoyed such a fine reputation as a touring machine. There was a slight bit of vibration when we reached speeds over 80 mph, however, this wasn't excessive. We feel that the R75/5 offers the rider the most comfortable arrangement of the three rider contact points—the handlebars, seat and footpegs. The seat is well padded and quite comfortable.

It is possible to log many, many hours without being fatigued because of engine or exhaust noise. This quiet performance has long been part of the BMW image. We do feel that the Bavarian mount is definitely a machine we would not hesitate climbing aboard for a three thousand mile excursion.

It is not unreasonble to call the BMW's design unorthodox. There is no question that there are some drawbacks to its powerplant. On the other hand we feel that the advantages definitely outweigh these drawbacks. For instance, it is the smoothest two cylinder motorcycle that we have ever ridden. Other manufacturers have had to go to three or even four cylinders to help hide and dampen the vibration so common to vertical engines. *Yamaha was the only other twin in the test group which also included two triples and two fours.*

We do feel that BMW will have to do something very shortly to retain its present position in the market. Design changes in the recent past have included a larger powerplant, more power, better brakes, a smoother ride and styling. It seems that it will be necessary for BMW to go through another metamorphosis of sorts before long. We feel this is necessary because of the efforts made and results achieved by the competition. The Japanese machines are impressive. No longer is BMW's finish quality that far superior (if at all) to other machines on the market today. Also it is now somewhat questionable whether the R75/5 is now the quietest and smoothest machine available. It is an expensive machine when considering the rapidly narrowing margin of excellence between the various models now available. Much of the mystique of the motorcycle seems to be based on its former superiority but today the situation appears to be one of equality as the East inexorably closes in on the West. ■

BMW R75/5

Engine type	four cycle opposed twin
Bore and stroke	82x70.6mm
Displacement	745cc
Compression ratio	9.0:1
Claimed horsepower	57 @ 6,400 rpm
Claimed torque	43 ft/lbs @ 5,000 rpm
Engine red lines @	6,400 rpm
Ignition	battery
Starting system	electric/kick, in neutral only
Carburetion	(2) 32mm Bing
Lubrication	wet sump
Type of transmission	constant mesh, four-speed
Clutch	dry single plate
Internal gear ratios	(1) 3.896, (2) 2.578, (3) 1.875, (4) 1.500
Braking performance, 35-0	39.6 ft.
Braking performance, 65-0	145 ft.
Quarter mile acceleration:	
Terminal speed	96.87 mph
Elapsed time	13.76 sec.
Acceleration, 40-60	3 sec.
Acceleration, 60-80	9.5 sec.
Actual speed at speedometer indicated 40	37.5 mph
Actual speed at speedometer indicated 70	68.0 mph
Indicated top speed	110 mph
Length	82.7 in.
Seat height	33.5 in.
Wheelbase	54.5 in.
Ground clearance	6.5 in.
Listed dry weight	423 lbs.
Actual weight, full tank of gas	475 lbs.
Front tire size	3.25x19 in.
Front brake type	internal expanding
Rear tire size	4.00x18 in.
Rear brake type	internal expanding
Air filtration	dry paper
Battery type	12V, 15AH
Fuel tank capacity	4.5 gal.
Fuel reserve	.5 gal.
Engine oil capacity	4.7 pt.
Gear box capacity	1.7 pt.
Drive shaft housing	.2 pt.
Rear wheel drive	.5 pt.
Front suspension	telescopic double damping
Rear suspension	adjustable spring over shock
Frame type	tubular double cradle
Steering damper type	friction
Colors	red, green, blue, black
Retail price, Los Angeles	$2095

DISTRIBUTORS
Butler & Smith
(West) 135 E. Stanley St.
Compton, Calif. 90220

(East) P.O. Box H
Norwood, N.J. 07648

AUNTIE MAUDE'S TEAM

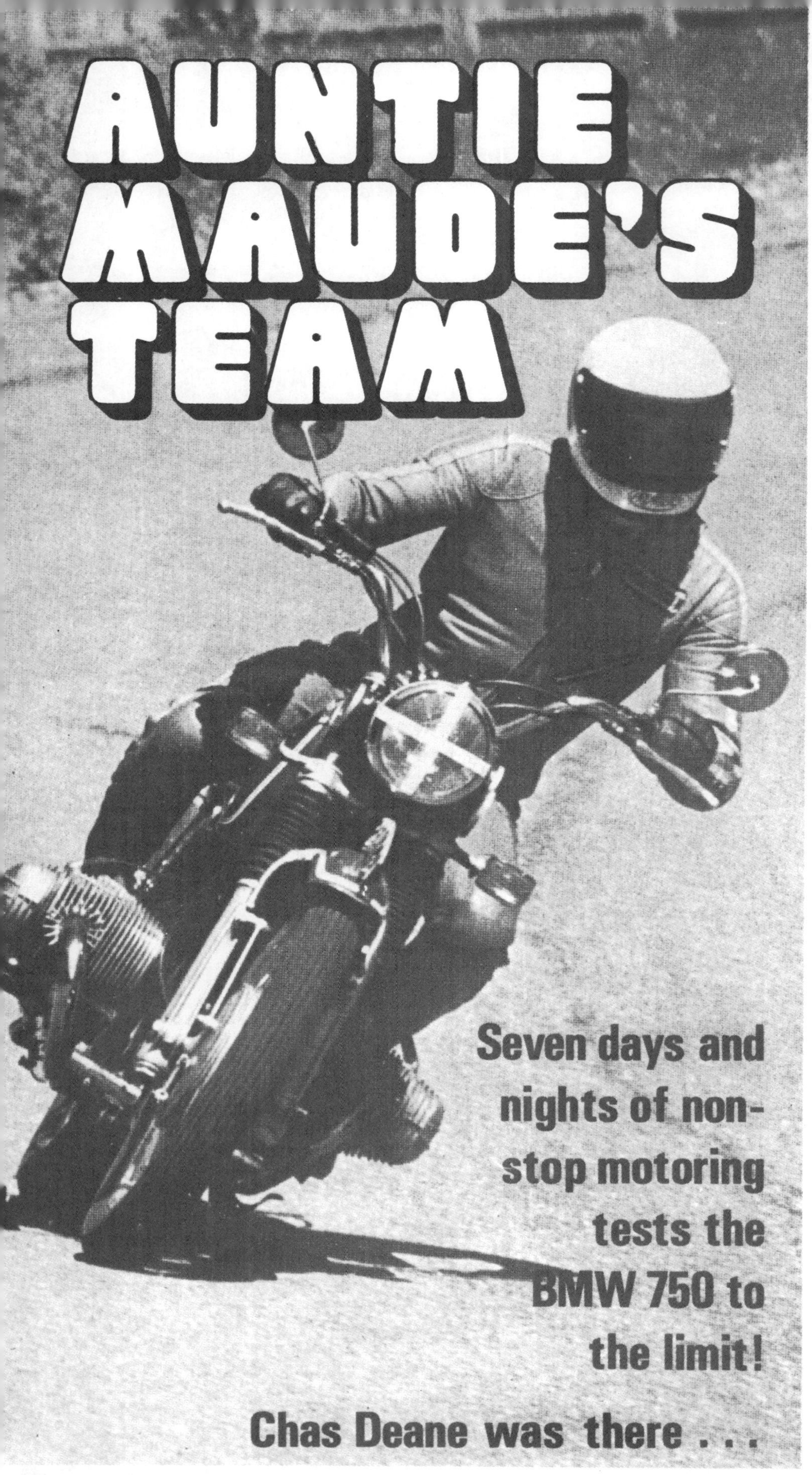

Seven days and nights of non-stop motoring tests the BMW 750 to the limit!

Chas Deane was there . . .

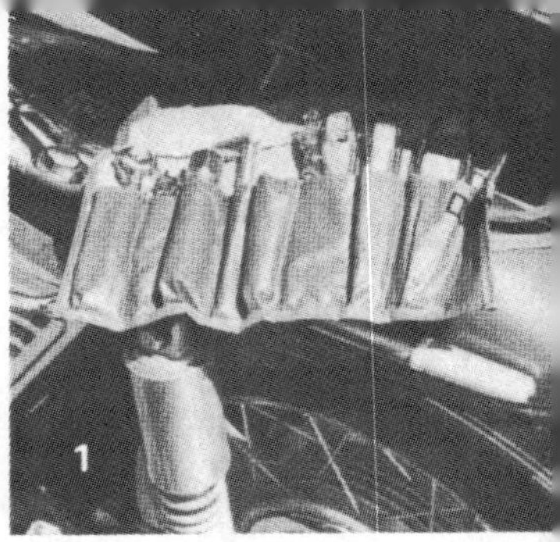

Battling along the Cronk-y-Voddy straight at 90 miles an hour in pouring rain isn't the best time or place to encounter a flock of sheep, but this was one of my more exciting moments on a BMW R75/5 in the Isle of Man during the Maude's Trophy attempt!

After seven days and nights of non-stop motorcycling, apart from refuelling and carrying out repairs caused by two unfortunate prangs, plus one burned out clutch, the two BMW machines covered a total of 16,658 miles over the toughest road circuit in the world, the TT Mountain Course.

Halfway through the week, the riders changed machines which meant that I was able to ride and compare both bikes during my 44-lap stint around the TT circuit. The machines were the latest off the BMW assembly line in Berlin and featured the new 2-inch longer rear swinging arm; a modification designed to improve straight-line high-speed stability. Other mods also included the fitting of Bosch quartz-halogen headlights, plus two ginormous spotlamps mounted on the crash-bars to help us through the night when there was mist up on the Mountain.

No special preparation had been allowed on the two BMWs, apart from the normal running-in and oil changes, plus adjustments to tappets and carburation. The engines had both been sealed by the German counterpart of the A-CU at the end of the assembly line and until the prangs on Monday morning, that

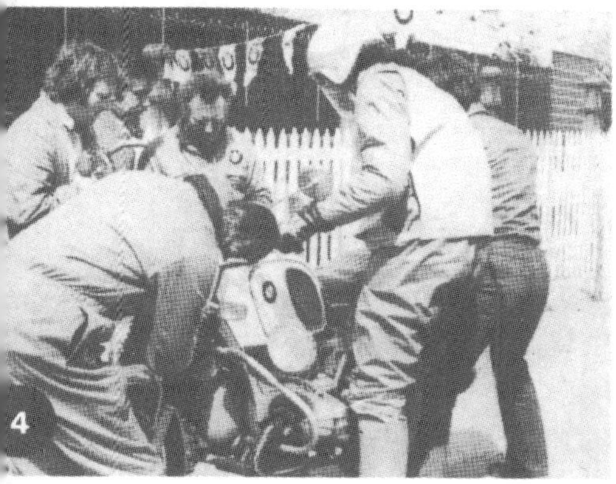

1. If the quality of a machine's tool kit is indicative of the rest of the bike then BMW deserve their reputation.

2. The big traditional motor — most distinctive thing on the 750 is the Bing CV carb.

3. The wide bars are less comfortable than the alternative flat type. With the exception of a heavy throttle action the controls are smooth and easy. We've given up making comments about the ignition switch.

4. Maudes' mechanics make some quick checks while Dai Davies gets ready to ride.

5. Tony Jefferies and Dave Minskip wait for the flag to start the Maudes' Trophy run.

was the way they had to stay.

Normally, when we road test a motorcycle, we keep the machine for about a fortnight to three weeks to make a complete study of it from all aspects, whether it be riding to and from work in London or carrying out performance tests on a closed circuit.

Obviously, with the idea of reliability in mind, my riding of the two BMWs on the Isle of Man meant a maximum rev limit of 5,500 rpm and a top speed of only 90 miles an hour, but for all this, most of the team of riders were lapping the 37¾-mile course in under 37 minutes to average over 60 miles an hour on roads open to the Manx motoring population.

In fact, the two machines were to average almost 50 miles an hour for the entire week's riding, including all refuelling stops and hours lost in repair work.

Anyway, returning to my point about road testing, I wasn't able to test either of the machines for maximum performance during the Maudes Trophy run and neither were we able to obtain standing quarter-mile times, but to compensate for this, on my return from the Island, BMW delivered a similar R75/5 to the office to fulfil these requirements.

Meanwhile, back on the Isle of Man, I can honestly say that the weather conditions were far from ideal for the Maudes Trophy run. I swear that at least 39 of my 44 laps ridden during that week were in rain varying from drizzle to torrential and that I had my own gloating giant of a black cloud following me around the course.

I personally ran into it many times at Guthrie's Memorial on the climb up the Mountain and continued to swim around in it, seeing no more than ten to 30 yards ahead of the BMW, for the next seven miles or more!

The surprising thing about the TT circuit was the amount of grip on the road no matter whether it was wet or dry and this was using Metzeler "hard" tyres which are supposed to give long life on the big BMW twins. In actual fact, they lasted about 3,500 miles on the Maude's Trophy run, which gives you some idea of how hard the machines were used and also the abrasiveness of the granite-chip road surface on the TT course.

Expensive

To be quite honest, I've never been an advocate of BMW motorcycles. I've always considered them much too expensive in relation to what you get for your pound notes.

Also, the technical specification of a horizontally-opposed, push-rod twin four-stroke with a four-speed gearbox doesn't appear all that grand in comparison with say the Honda 750 four-cylinder, overhead camshaft, five-speed unit. Admittedly, the shaft-drive is nice and keeps the bike free from grease splashes and frequent chain adjustment, but who really wants to pay an extra £300 for the privilege of not having to adjust a chain every so often?

But, having travelled over 1,500 miles in a week under the most appalling weather conditions, I am now prepared to admit that here is a convert to the shaft-driven twin from Germany.

I also admit that I still don't like that clonky gearchange, but as for the rest of the bike, well I reckon that very few other machines could have taken the punishment which we dished out to those two 750s over the TT circuit.

The Maudes Trophy attempt started at midday on Thursday, May 3rd, with TT ace Tony Jeffries and London speedcop Dave Minskip being waved off the TT start line on the Glencrutchery Road by Geoff Duke and the Mayor of Douglas.

My first stint at the handlebars was to be later that evening, just as the sun was setting and the locals were making their ways to the Manx pubs. I'd circulated the course a few times during previous visits to the Island, but it wasn't till I started on that very first lap that I realised how little I knew about the 37¾-mile circuit.

I was shutting off and changing down gear when I should have been opening up and the fact that it was bucketting down with rain didn't inspire confidence on the very wet road surface.

Mist on the Mountain added even more to the misery of trying to maintain a reasonable average speed. My first lap took 48 minutes, which was to prove my slowest of the week. By this time, I'd begun to realise that there was considerably more grip on the wet roads than seemed or appeared possible. Only where the circuit had been resurfaced and that white plastic centre of the road lining material had been used, was there any sign of tyres breaking away.

I was riding machine B at this time and accompanying me on machine A was freelance journalist and commentator, Alan Robinson. On the corners, presumably to keep the bike as upright as possible, he used the "slide-out-of-saddle" technique . . . quite a sight when you are following him through the nadgery bends at Laurel Bank. Alan has ridden in the TT on a Honda 50 when they used to run that class on the Island and obviously knew a fair amount more about the circuit than I, particularly at sections like Doran's Bend and the 13th Milestone, where you have to know exactly where the road is going to ensure you're in the right gear travelling at the right speed.

Needless to say, I learned a fantastic amount about that TT circuit on those first four laps, during which time we reduced our lap times down to around the 40-minute mark. Not fast, but at least it was safe and within the required 50 miles an hour average which had to be set to achieve the target figure of 17,000 miles for the two bikes in a week.

I also discovered a lot about the new 750 BMW. It handles a real treat on long, fast swooping bends and although a little on the heavy side for flip-flop nadgery corners, the ground clearance allows one to crank the machine to unbelievable limits. In fact, the first thing that touches the ground is either the small tang on the centre stand or the prop-stand, I'm not sure which!

a year's riding in one week...

The brakes? Well, we never had to make use of them to any real extent, except for the fast approach to Ballacraine, where you turn sharp right at the hotel, and on the drop down to Creg-ny-Baa. At both of these points, I thought a little less effort and a better result is needed from the BMW stoppers, but at least water didn't fill the drums during their long periods of inactivity.

The motor? Smooth and sweet as a nut until the two prangs took the edge off both of them. The A machine was cartwheeled down the road at about 70 miles an hour when Dai Davies touched the wall coming out of Brandish Corner. Result? Bent frame, smashed rocker cover and bent rocker spindle mountings, plus a complete front fork write-off. It took the BMW team of mechanics, Ted Gatward, Dave Houghton, Keith Jones, Jim Mason, Les Tyler, led by Alberto Criscuolo only four hours to collect the bike, strip and rebuild it and have it back on the road.

From this moment on, the tappets really rattled, which is not surprising considering the rocker mountings were bent. I was amazed that the bike ever ran again and completed its seven-day stint. It says a great deal for the robustness of the engine castings.

Anyway, as I have said, the motor ran as sweet as a nut up to the moment of impact. It was very tractable and apart from the tic-toc swaying motion felt through the entire machine on tickover, the faster the engine revved the smoother it seemed to run. Even the handlebar rear view mirrors could be used at 80 mph-plus!

The riding position with the large, deeply-padded dual-seat was very comfortable, although by the end of the week the foam interior had been compressed to the shape of the rather large, six-foot plus form of policeman Dave Minskip, slightly smaller but also heavy Peter Kelly of the *Northern Echo* and no lightweight, TT Tony Jeffries!

Riding controls were medium to light in operation, except that I found after about two hours in the saddle my right wrist was starting to ache due to the slightly heavy throttle twistgrip.

Fuel consumption during the run varied between 40 and 50 miles per gallon, depending upon who was at the handlebars and how hard they were riding.

I handed B bike over to Bruce Preston of the BMF and Alan Robinson swapped with John Nutting of *Motorcycle*. Their first question as we splashed into the pits was: "What's it like on the Mountain?" Invariably our reply was a simple: "Bloody awful!"

Fortunately, there were one or two more lighthearted moments to brighten the dim weather like when Alan and I were cranking out of the corner at Kate's Cottage and found a lively little hare in our path. I swear he out-accelerated both of us down the hill with puffs of smoke whipping up from his high-speed, ultra-grip rear paws!

Then there was the darling old lady who pulled out of the small side road adjoining Ballaugh Bridge, and stopped in the centre of the road just as I was about to leap the bridge. Fortunately, I was wearing my face mask and was rendered speechless as I screeched the BMW to a halt about 18 inches from the driver's door.

The other riders in the team; Mick Hemmings and Mike Nicks of *Motorcycle News*, John McDermott of *Motorcyclist Illustrated*, Alan Killip, Dave Wood and Brian Kewley, all local lads, had many similar tales to relate. Even the team managers, Ken Heanes and Alec Smith took a turn at the bars and actually did the last four laps of the run in the rain to end what for me can only be described as a memorable experience.

The only regrettable thing on the attempt to win the Maudes Trophy were the two unfortunate prangs. The second happened when Mick Hemmings found a lorry in his path waiting to turn right as he was riding round a left hander. The resulting prang caused the motor to over-rev and bend a valve, which later had to be repaired, plus all the front end damage.

There's no doubt about it, the two BMW 750s took more punishment in one week than the average rider would give his machine in a year or more. Is the attempt worthy of the award of the Maudes Trophy?

Only the ACU can answer that when they sit to discuss the effort in committee next month. Obviously, I am biased because I was one of the crazy, mixed-up masochistic riders who suffered at the hands of the Manx weather for seven days and nights. But whatever the outcome, I'm pleased to say I was there, thanks to Douglass Austin of BMW, who organised this fantastic effort.

Charles Deane

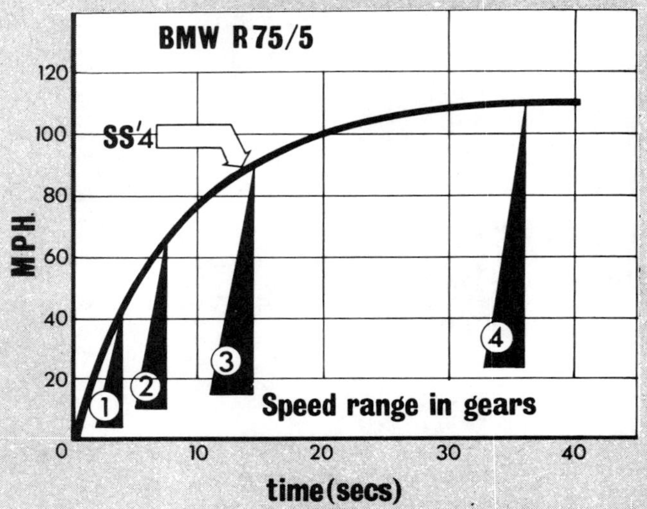

SPECIFICATION

braking distance from 30 mph	28ft.
fuel consumption	40 to 50 mpg
engine type	horizontally opposed ohv twin
displacement	745 cc
bore x stroke	82 x 70.6 mm
compression ratio	9.0:1
claimed output	57 bhp at 6,400 rpm 36.9 lb. ft. at 5,400 rpm
lubrication	wet sump, high pressure pump
carburettor	twin Bing CV 32 mm
ignition system	twin Bosch coils, contact breaker
lighting system	Bosch QH headlight 12 volt
charging system	Alternator/Bosch regulator
clutch	single-plate, diaphragm spring
gear ratios, internal	
first	3.896:1
second	2.578:1
third	1.875:1
fourth	1.50:1
final drive	shaft with spiral bevel pinion 2.91:1 reduction
mph/1,000 rpm in top	16 mph
fuel tank capacity	6.3 gals.
oil tank capacity	4.75 pints
tyres, front	3.25 x 19 Metzeler
rear	4.00 x 18 Metzeler
brakes, front	7.9 in. twin-leading-shoe
rear	7.9 in. single-leading-shoe
wheelbase	54.5 in.
ground clearance	6.5 in.
seat height	33.5 in.
dry weight	419 lbs.
test weight	463 lbs. including oil and fuel
suspension, front	telescopic, hydraulically damped
rear	s/a, 3-position dampers
trail	3.35 in.
speedo error	accurate at 50 mph
tacho error	accurate at 5,000 rpm
tool kit	excellent
original equipment	electric starter, flashing indicators, mirrors, headlamp flasher
Importer	BMW Concessionaires Ltd., 91 Great West Road, Brentford, TW8 9ED.

DREAM COME TRUE

BMW introduce the super twin to match the multis. Charles Deane rides and reports on the new R90S sportster from Munich . . .

What could be better than a BMW 750 twin? Answer . . . the brand new R90S twin with a capacity of 900ccs producing a very nice 65 DIN brake-horse-power!

Couple this with a new five-speed gearbox, double disc hydraulic front brakes, plus complete restyling and you have a dream come true . . . a superbike to "out-superbike" the current range of glamorous multi-cylinder machines at present on the two-wheeler market.

My flight into the realms of pure fact and no fantasy began at London airport where a BEA Trident winged me to Munich and the headquarters of BMW. I had heard that there was a new bike awaiting announcement, but I never dreamed that such a revolution had taken place within the research and development department at BMW. All the items which had received constant knocking from the motorcycle press . . . the clonky gearbox, the outdated front drum brake, the lack of horses in comparison with other large motorcycles . . . BMW had listened, learned, developed and produced the answer.

The R90S is obviously going to cost a lot of bread, but for your money you get what must be one of the most exclusive and exciting machines on the road.

It may not have four or six cylinders, but as BMW say: "If you can produce the power and performance plus reliability with a comparatively simple push-rod twin of established and proven design, what's the point in making things more complicated just for the sake of complication?"

And I must admit that I am about to agree with them, particularly after having ridden practically every one of the multi-cylinder machines around.

The view which most other riders and motorists must see of the new R90S BMW! New silencers are as efficient as ever to stop noise

The twin-disc brake had brought the sports BMW bang up to date and provides superb stopping. A squeal at low speed was annoying

Who needs a flight check before take-off? This handlebar layout is standard with sports screen and full instrumentation as seen

The Dell'Orto carburettors are of "concentric" design and have a 1½ inch choke. They help add to the performance of the twin

Standard BMW rear-end with the clean, fuss-free shaft drive which passes through the right-hand side of the swinging arm

Where it came from! Dave Potter racing the prototype of the new BMW 900 sportster in the 24-hour endurance race in Barcelona . . .

Outwardly, that is engine-wise, the BMW doesn't look any different to the R50-60-75 models. That is because, apart from an increase in bore size, it isn't!

However, carburation is very different because the R90S uses two, 1½-inch Dell'Orto "concentric"-type units, an immediate giveaway when compared with the Bing CV carbs used on the R75/5 and standard R90 models.

The five-speed gearbox must be considered one of the major modifications or introductions for the 900 cc BMWs, because it allows closer ratios to be used and this in turn removes much of the clunkiness found on the four-speed box.

I would like to report that all sign of gearbox clunking has gone, but although greatly improved over the four-speed gearbox, the five-speed unit still retains a slight tendency to clunk-click every trip!

Apart from a lengthened swinging arm unit, the rear transmission is normal BMW with its foolproof shaft-drive which not only keeps the rear of the machine immaculately clean and free of oil, but unlike chains on other large capacity bikes, doesn't constantly need adjustment. A very good point.

At the front end, the superb two-way damped telescopic front forks carry a 19-inch wheel which is now equipped with double-disc hydraulic brakes.

To say they are efficient is putting it mildly . . . they are fabulous in the way they will pull 474 pounds of machinery, plus rider, to a standstill from any speed. The only annoying thing was a tendency for them to squeal when being applied gently. The brakes are also unusual in that the master cylinder isn't incorporated in the handlebar lever as on the vast majority of motorcycle hydraulic brakes.

The master cylinder and fluid reservoir is under the fuel tank and operated by cable from the handlebar lever. Obviously, it is not easy to give a visual check on fluid level without first removing the fuel tank, so BMW being what they are, have incorporated a warning light in the system which lights up on the dash panel in front of the rider should the fluid reach crisis level.

There are similar warning lights for oil pressure, battery charging circuit, ignition and flashing indicators; add to this the voltmeter, rev counter, speedo and electric clock and you have quite an impressive dash panel all compactly mounted beneath a half-fairing sports screen.

A hydraulic steering damper is another of the refinements added to the R90S, not that it was necessary under normal road conditions. And, finally, we come back to all the standard BMW items such as flashing indicators, electric starter, stoplights operating from front and rear brakes, headlight flasher and wing mirrors that can really be used due to lack of vibration from the motor.

Nothing need be said about the immaculate BMW finish except that it is up to the normal high standard and the unusual two-tone paintwork unique to the R90S only adds to the appeal of this superb machine.

Introduction over, there came the one thing all journalists present had been waiting for . . . the chance to actually ride the new 900 sports twin. Due to my missing the early flight to Munich, I was one of the last to arrive for the test day, but it had its compensation because whereas the other journalists were all restricted to a few laps on a test track, I had the bike all to myself the following day for a marvellous run out into the countryside around Munich. Autobahns, open roads and country lanes . . . I was able to revel in the pleasures of powering the gentle giant along at over the "ton" tucked down behind the sports screen on the autobahn, pinning it down and squirting it through the nadgery corners in the Bavarian countryside and "dragging" it away from traffic lights . . . a few oh-so-short hours of delight for a motorcycle journalist.

So, how does the R90S really perform?

Starting? Say no more . . . just tickle the carbs, turn on the "real" ignition switch and press the starter button. I say real ignition switch because at long last they have got rid of that switch which could be operated by a six-inch nail! I should think that in itself took a revolution in the BMW plant!

Anyway, with the motor ticking quietly like a watch, the clutch lever was pulled in and although slightly on the heavy side, it was smooth in travel.

Now providing one hasn't made the mistake of blipping the throttle before pulling in the clutch, it is possible to select first hear without so much as a click, let alone a clunk.

The next stage is moving off and this depends on what mood you're in; if you want to blast off for a standing start quarter mile of around thirteen seconds, then give a big handful on the twist-grip, drop the clutch and hang on! But the BMW rider is sedate and with a mere whiff of gas, the R90S will burble forward as smooth as a hearse going to a funeral.

The great thing about BMW is the lack of fuss from the bike, be-

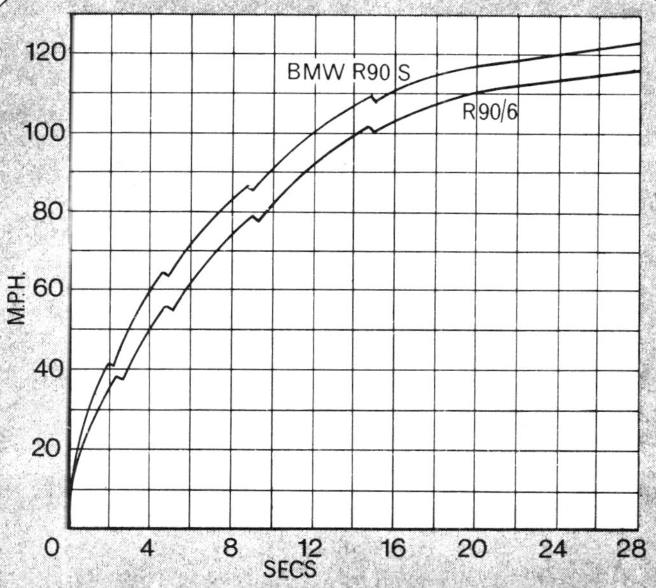

SPECIFICATION

braking distance from 30 mph	not tested
fuel consumption	not tested
engine type	horiz. opposed twin
displacement	898 cm³
bore x stroke	3.54 in. x 2.78 in.
compression ratio	9.5:1
claimed output	65 bhp (DIN) at 7,000 rpm
lubrication	wet sump, pressure feed to bearings
carburettor	Dell 'Orto 1½ in. PHM
ignition system	twin coil C.C. unit
lighting system	12-volt Bosch 280 watt alternator
charging system	via solid-state voltage regulator
clutch	single-plate diaphragm
gear ratios, overall	4.40, 2.86, 2.07, 1.67, 1.50:1
primary drive	layshaft reduction 1.50:1
secondary drive	final drive 3.00:1
gearbox sprockets	not applicable
wheel sprockets	not applicable
mph/1000 rpm in top	not tested
fuel tank capacity	5.83 gals inc. reserve (.55 gal)
oil tank capacity	not available
tyres, front	3.25 x 19
rear	4.00 x 18
brakes, front	twin hydraulic disc 10.4 in.
rear	s.l.s. full width hub 7.87 in.
wheelbase	57.7 in.
ground clearance	7 in.
seat height	32.3 in.
dry weight	not available
test weight	474 lbs
front/rear ratio	not available
suspension, front	telescopic forks two-way damped
rear	swinging arm
castor	not available
trail	not available
speedo error	not tested
tacho error	not tested
tool kit	excellent
parts prices:	expensive
list price	not available
original equipment	electric starter, flashing indicators, wing mirrors, headlamp flasher, electric clock, tachometer, voltmeter
Manufacturer	BMW Motorcycles GmbH, Munchen, Germany.

Dream come true!

cause both induction and exhaust are so well silenced to pass the very strict German noise regulations, that even when on full song there isn't the slightest hint of ear-drum bursting crackle from the motor.

The red line on the tachometer starts at 6,750 rpm and if you're giving the bike some stick in first and second gear, the indicator needle positively darts towards the red line.

Power starts to pour in at about 3,500 rpm and by the time 5,000 rpm is indicated, there's a very hefty, silent hand pushing you hard in the back.

With the closer ratio five-speed box, the engine stays in the power band right the way through to top gear, so there is no slowness in gearchanging or acceleration. The "ton" is reached in approximately twelve-seconds and a maximum speed of 122 mph can be achieved while in a sitting up position. This can be improved on by crouching down with your head about level with the sports screen and if you are wearing racing leathers and really "tank it", a true top speed of over 130 mph can be obtained.

Gone is the BMW image of building bikes for the middle-aged touring man. This is the bike for the full-blooded young man with cash in his pocket who wants the ultimate in two-wheeler comfort and performance.

The R90S is the motorcycle for the young executive who runs a Lotus Elan, Alfa Romeo or the like, but who is fed up with traffic jams and longs for the open road.

Stopping from high-speed has never been one of the BMWs best points, but they have put this right on their sports 900. The twin discs can be made to hop the front wheel and squeal the tyre at almost any speed and it is necessary to treat them with a great deal of respect in the wet or greasy, town roads. They will lock the front wheel too easily under these circumstances, so you have to ride accordingly.

Handling on the R90S is very good, but 474 pounds is still a lot of weight to throw into a long, fast corner. Therefore, when I did get a slight wallowing after hitting a bump on a bend at about 90 mph, I wasn't really surprised. Mind you, it wasn't worrying and at no time did I feel that things were getting out of control.

On tight corners and nadgery bends, it's surprising how one can throw the BMW around, in spite of what people say about the torque reaction of the motor when opening up the throttle. Theoretically, the bike is supposed to wind itself around the motor as you open the throttle, this can easily be experienced by blipping the throttle while stationary, but in practise it has no really noticeable effect on handling.

There's no doubt about it, I really enjoyed riding that R90S and I'm longing for the day when we get one for a full week or fortnight for test. As it was, I wasn't able to carry out fuel consumption checks or night riding (lighting) tests.

I know the machine was very comfortable to ride with its large, deeply-padded dual seat and all controls perfectly set to hand and foot.

Vibrations could be felt low down the rev range of the motor, with a period felt through the handlebars at about 4,500 rpm, but after that the motor simply went faster and became smoother!

So, what about the price? As yet no announcement has been made in Britain, but with the R75/5 costing approximately £1,150 it is fairly obvious that with all its refinements, the R90S is going to cost considerable more. I'd guess it to be around £1,550 to £1,600 for the lot including air in the tyres!

Is it worth it? Who can say? But is a Rolls Royce worth the £8,000 or so people pay? All I know is that it is one hell of a good motorcycle. I wish I could afford to buy one.

Fifty-Five Cubic Inches of Guaranteed Exclusivity

By Bob Atkinson

RIDING BMW'S SENSATIONAL R90S

BMW HAS definitely got a problem. Look at it this way. Our American image and their European image just don't coincide. When we think of a BMW we think touring, saddlebags, a fairing maybe, and sedate performance to go with it. Sounds like a pack mule doesn't it? And, well, pack mules just aren't that exciting.

Close your eyes for a minute and imagine yourself in Europe. Anywhere will do, as long as it's not too far from a major highway. Sooner or later a BMW rider will flash by. I say flash, because he won't be going slow, picking flowers along the side of the road. He'll be wearing either full leathers or a rain suit, and his bike will have a 5-gal. tank, narrow bars, and a skimpy luggage rack or no rack at all. Their image, quite simply put, is a combination of sport and affluency...kind of a cross between Don Juan and the Red Baron.

As for performance, Americans and Europeans are at odds again. We think in terms of the quarter-mile, and the BMW marque has never been a threat here. But the European regards performance as getting from point A to point B in the shortest possible time. A typical illustration would be gassing it up in, say, Munich, Germany, and letting off in the vicinity of Paris, France some hours later! Some of the braver pilots will make the journey hovering in the 6500 to 7000 rpm range in top cog, and that is flat hauling.

Now, for hauling even faster and with a considerable amount of style, BMW has the R90S. More cubic inches. More class. Fantastic.

Appearance wise, the R90S is like no other BMW ever made. It isn't really a cafe racer, but it definitely leans in that direction. A rounded tail section tastefully houses the taillight and blends in with a lower seat with a better contour than found on other BMWs. The tank is

a 6-odd-gallon affair which narrows appropriately at the rear, allowing the rider to become one with the machine. Then there's a partial fairing, that adds a lot more style than protection. Still, it's something to tuck in behind as speed increases. Completing the package are dual disc brakes (the master cylinder is located under the gas tank) and rubber sliders instead of full gaiters on the front suspension.

It's much like any other cafe racer... until you add the paint. BMW has chosen a fogged grey with a tonal gradation that goes all the way to black. If anything, it makes the R90S look sleeker than it really is. Pretty crafty, those Germans.

Appearance isn't the only ingredient necessary to change an image. Performance weighs heavily, and BMW has done something about that, too. When you straddle an R90S, the first thing that catches your eye is a new instrument group. There is now a separate tachometer and speedometer. Just above these are two smaller gauges—a clock and an ammeter. It's kind of like the dash on a small GT car. Push the electric start button next to the right hand grip and the 900 springs instantly to life.

Surprisingly, the 900cc S idles smoothly, in spite of its slightly more radical camshaft and larger carburetors. Both the 750 and the new 900 T have constant velocity 32mm Bings. The S comes with 38mm slide type Dellorto PHMs. When you let out the clutch for a normal take off, the S feels about like a standard 750. There's a little less torque right off the bottom than the 900 T has. But, get the S engine up to around 4000 and punch it and there is considerably more steam, right up to the 7000 rpm limit.

I got a chance to run the S hard on BMWs test track outside of Munich and there, BeeEms finest is a convincing,

Photography: Bob Atkinson

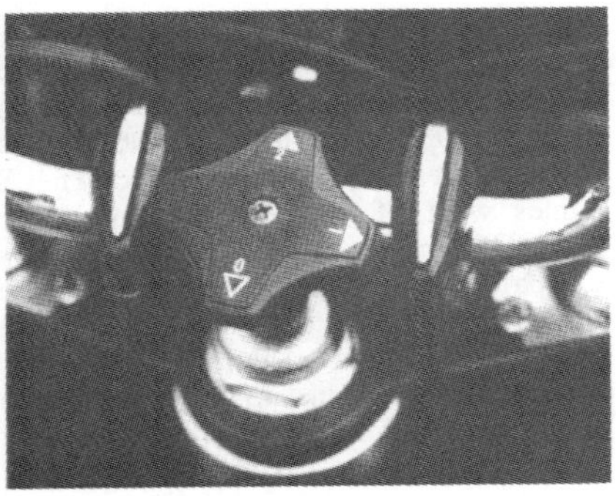

stable performer. Out of any of the numerous tight turns, the S accelerates smartly, thanks to accelerator pumps in the new carburetors. For the most part the engine will respond instantly, but the rider's task is hampered by a throttle that is not quick enough for precise control.

There's a left at the end of the front straight that's a true test for brakes. On the S, two fingers and moderate pull will let you squeal the front tire. Feather in the rear for more control and the 474-lb. machine halts quickly, every time.

You don't really pitch a BeeEm into a turn, rather, you lean it over with considerable body english, especially if it's a near 90-degree turn at the end of a straight. If you persist (I did), the side stand will ground in turns like this, but nothing comes of it as the stand is too flexible to lift the rear wheel off the pavement.

Next comes a downhill hairpin-right. You touch the brakes, apply body english (lots this time) to shift the bike over center and gas it down to a hairpin left. In most of these corners I was banked over far enough to grind the bottom of the foot pegs away. Until I got a little carried away and lifted the rear wheel off in one turn, the fact that I was dragging didn't enter my mind. The R90S is that stable.

The remainder of the handling course is a series of sweepers and a jump that can be negotiated in a variety of combinations. You've got to watch it a bit over the jump. If you come down with the power on, the BeeEm gets a little squirrelly. In any of the sweepers, it was stable and steered well, right up to the point the rear end began to drift.

As for grinding holes in valve covers from dragging them in turns, don't worry about it. It just doesn't happen very often, and you have to be leaned way over when it does.

There wasn't really room on the handling track to get into fifth (all 1974 BeeEms are five-speeds), but I did change a lot of gears, and happily, the famed BMW clunk is not nearly as pronounced as it has been in the past. This isn't the result of a new cush-drive setup. Rather, it is a byproduct of closer gear spacing which allows engine and transmission speeds to synchronize a little faster.

Technically, the R90S is similar to the rest of the BeeEm line. The engine is an overhead valve, opposed Twin with the camshaft positioned below the forged steel crankshaft. The cam is driven by a double-row chain with a tensioner unit, which does away with chain slack. The crank is similar to the 750 (both have the same stroke), but it is balanced differently to work with the larger pistons in the 900. The barrels, naturally, feature a bigger bore, but are still made from aluminum alloy with a cast-in steel liner.

Compression is up to 9.5:1 from 9.0, and as mentioned earlier, there are larger carburetors with accelerator pumps. All together, the setup is good for 67 bhp (DIN) or 77 bhp (SAE) at 7000 rpm.

It's not the most powerful motorcycle on the road, but it'll go fast enough to scare anybody. And it will do it mile after mile after mile.

As for price...we don't know yet, but it will be expensive. Class, though, has never been cheap!

BMW R90S

Suspension, front	telescopic fork
Suspension, rear	swinging arm
Tire, front	3.25-19
Tire, rear	4.00-18
Engine, type	ohv opposed Twin
Bore x stroke, in., mm	3.54 x 2.78, 90 x 70.6
Piston displacement, cu. in., cc	54.8, 898
Compression ratio	9.5:1
Claimed bhp @ rpm	67 (DIN), 77 (SAE) @ 7000
Piston speed @ rpm ft./min.	3243 @ 7000
Carburetion	(2) 38mm Dellorto PHM
Ignition	battery and coil
Oil system	gear pump, wet sump
Oil capacity, pt.	4.75
Fuel capacity, U.S. gal.	6.3
Recommended fuel	premium
Starting system	electric, kick

Air filtration	treated paper
Clutch	single-plate, dry
Primary drive	none
Final drive	shaft and bevel gear
Gear ratios, overall:1	
5th	4.50
4th	5.01
3rd	6.21
2nd	8.58
1st	13.20
Wheelbase, in.	57.7
Seat height, in.	32.3
Handlebar width, in.	24.8
Ground clearance, in.	8.5
Curb weight (w/full-tank fuel), lb.	474

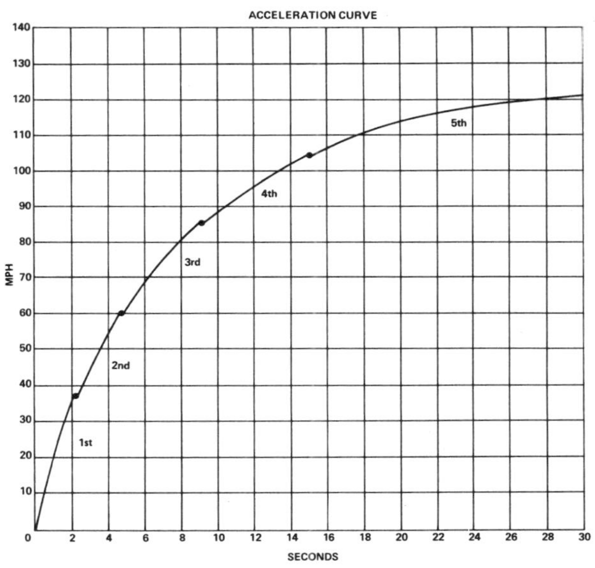

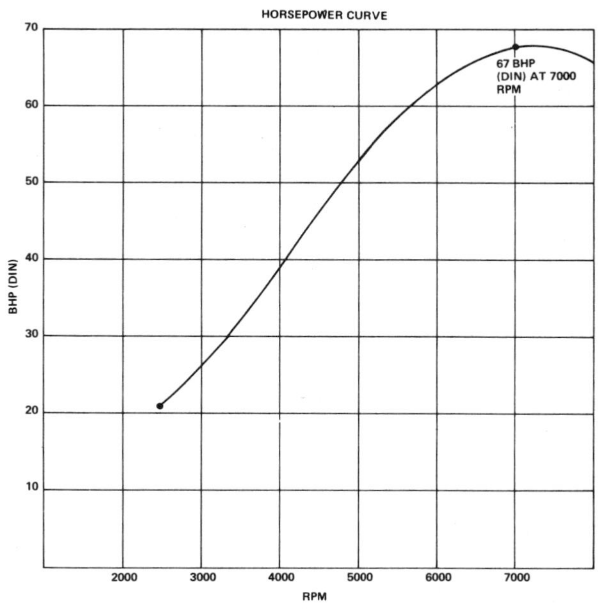

FORGET ALL YOU'VE EVER READ ABOUT BMW TWINS

Test: John McDermott
Photography: Vic Barnes/John McDermott

THE Marketing Department in Munich has explained the matt black handlebar levers as being designed to cut out dazzle. Cobblers.

Without doing any research on the subject whatsoever I shall stick my neck out and declare that the National Health Service and BUPA will hardly notice the fall in demand for hospital beds by bikers blinded by glare from this source enough to stack their plot into oncoming traffic, so if BMW are pitching for the Nobel Prize for Services to Humanity, they ain't gonna get it on that basis.

They may, however, be in the running for the Lombard Finance New Business plaque, for the price of the thing coupled with its desirability may well encourage a fair number of people to mortgage themselves up to the eyeballs in order to buy one.

Make no bones about it, this mother is not for your common or garden biker. Anyone who can afford eighteen hundred notes for a motorcycle is anything but common and probably employs a gardener or three anyway.

Few of us — and you can include me and everyone I know in that exception — can afford such a bike, but don't dismiss the R90S as not being worth the money. Certain things cannot be costed in pennies or pounds, and whatever these things might be, the R90S has three each of 'em. The word "charisma" was coined for just this dilemma: if it makes you smile and nod your head in awe and approval, it has charisma. And when you ride one of these Silversmoke coloured flat twins, the whole world smiles, nods and awes until you get sick to the back teeth of smiling back.

BMW have their own variation of the old adage "You pays your money . . . . That you most certainly do, but you take no choice but theirs: this year it's called Silversmoke, silver in parts, it sort of smudges into bronze and brown around all the edges. Not everyone's choice but then the manufacturer has made it easy for you by giving you none.

Earlier this year the factory took me to Germany to try the thing out and I was enthralled by its performance even though I rode it for a very short time. A bit later BMW (GB) flew me to the South of France for a longer ride and I enjoyed it even more. My plan was to fly down to Marseille to pick up the bike and then ride it home, but when ten bikes go into France in one truck and on one consignment note, then ten — not nine — must go out the same way, so my 800-mile test was blown to the winds.

Nevertheless, a day's riding in France gave me the chance of screwing it up the Autoroute, plodding around villages and gushing up and down mini-mountains in company with other motorcycle journalists. Enough to get a better impression of the machine than I had already. Upon returning to the UK I had another one for a week.

Top whack down the Autoroute? Would you believe 129mph on the clock with a slight three-quarter sideways breeze? This speed was held for six whole minutes until the turn off for the lunch break appeared and I wound her down for the long, loopy sweep off the motorway and down to the cash desk on the exit. That reading was obtained by tucking everything as tight as I could manage into the metalwork and assuming a most ungainly posture that would not be comfortable for any length of time. Sitting half up in a normal fast touring stance cut the needle back to just over 122mph and rolling the twistgrip off a shade in deference to the pile of alloy rushing up and down inside the cases gave an acceptable cruising speed of about 117mph. Even at maximum the bike felt good and I was surprised how unfast 129mph seemed. I suppose that's normal, everything is relative until something goes wrong and then you just *know* you're tramping on.

Riding position is superb. The short, almost flat handlebar allows one to lean forward neatly into the cockpit across the long, slim tank and the footpegs complement this position absolutely. The seat, however, is a shade too thin and after a few hours in the saddle one can feel the

This one is different

sharp sides of the glass fibre pan poking through. First-time BMW jockeys might find the carbs bashing their feet as I did a few years back, but one learns very quickly the cure for sharp stabbing pains in the foot.

One thing very noticeable by its absence was the lack of engine braking. Strangely, it varied in great dollops at different rpm readings, seemingly normal up to 3,500 with bags of braking, it got less and less until the motor reached 5,000. Then there was nothing. When you rolled her off, she just kept rolling along. The "S" motor also needs a fair amount of gearbox stirring to keep her whipped up and flying as throttle response was markedly slow at low revs. Turning the tap on hard from low rpm produced a rumbling knock from within and a mediocre accelerative force, the kind of reaction one would expect from a 12:1 compression ratio on three star fuel. That may be an unfair criticism as the R90/6 is the motor for you if you just want to indulge in normal riding, so let's say this was an observation rather than a dig. After all, the "S" model is a sports tourer designed for adrenalin-pumping long haul sprinting.

The lack of engine braking on downhill and flat swervery gave a proportionately greater task to the brakes, for even cog swapping was not enough to pull the bike down to the right pace for some of the bends encountered in the hills outside Avignon. The brake lights must have been showing for fairly long spells on these roads and that is one thing that I did not like about the bike. Flashing stoplamps more often than not indicate that the rider is not entirely in charge of the machine — or that he lacks the skill of reading the road ahead and making the appropriate commands to the machine in good time. Maybe in time I'll be able to get it all together with this bike but I doubt it.

The cockpit is neat in the extreme. All the /5 switchgear has been retained, there is the usual praiseworthy lack of stringy cables protruding from the controls and the speedo/tacho/warning light cluster is easy to read and must have been designed by an artist. The Voltmeter and time clock in the fairing are clear and precise, and the incorporation of the last item is a feather in the cap of the Munich team. The small fairing itself is good too, but had it been just a shade — maybe two inches — higher, it would have been excellent. As it is, it tends to direct the deflected wind into the forehead and gives the neck muscles a bit of a workout, an exercise the head could well do without since the introduction of heavier helmets.

There's not much to say about the engine itself as a piece of visual engineering for it differs little from the totally encased and protruding lump it has always been, but I have a small bitch on the subject of the carburettors.

Big throated Dell'Orto jobs, they need tickling every time the bike is started whether cold or warm, certainly if one wants an instant fire-up to save the loading on the battery. But the tickler button is so short it allows fuel to spew over the operating finger, and petrol soaked leather and smelly fingers aren't at all nice. The long black plastic button on the more familiar Bing carb would be the answer. Hard as it is to level any great criticisms at the R90S, I don't think I'm being small minded in pointing the finger at this item, after all, BMW were among the first to levitate above the dirty fingernail, oil on the pavement image of old.

A Bee Emmist reeking of high octane fuel like he's just done a hosepipe job on a parked car is not quite the image the Berlin and Brentford publicists are aiming for. One can see the haughty noses at the Hilton sniffing the air as our groovy young hero escorts his date to the bar. Chelsea Bridge, OK but not Park Lane.

Vibration on the familiar flat twin is a fairly well known quantity by now, there is a touch of it at low rpm but this can be seen rather than felt. Below 3,200rpm the rear view mirrors jazz around enough to give double vision but a gloved pair of hands remain oblivious to this. Feet likewise are insulated from any buzzing, and over 3,200rpm everything smoothes out. The left hand mirror of the machine I had was not tightened properly and this resulted in the thing swinging back with the vibration and clattering against the fairing. With the assistance of the ample toolkit housed under the seat, the problem was quickly overcome.

The seat is lockable and hinges across to give access to the tool compartment, a decently large space which contains the toolroll and a copy of the BMW Europa service book, an excellent publication with road and town maps and a list of all service dealers throughout the continent.

The seat itself I have already commented on with regard to comfort, but I should add that it is good and long, sensibly wide and the grab rail at the back is fine for shock-cording bits onto the seat, but not very good as a grab rail. Only at

This one is different

the back does it have enough room to facilitate gloved fingers to hook around it.

The glass fibre nacelle at the rear of the seat acts as lid to a second under-saddle container box which would carry a light coverall or rainsuit quite comfortably. A tyre pump (good to see this item back as standard equipment) sits inside the rear frame loop and the flip-over seat is lockable with the ignition key thus keeping all contents reasonably safe and secure. A helmet hook is welded onto the frame loop and this is also protected once the seat is locked.

Passenger comfort, with a larger helping of foam padding than the pilot has, is good but the footpegs are a little high for leggy people despite their adjustability. The restyled silencers are maybe a bit better looking than the old ones, but they follow the same sweep upward towards the rear and time and again I have seen melted plastic heels splodged on the chromework just behind the footpegs. Ridiog boots are still made mostly with leather heels but ladies footwear nowadays seem to be entirely equipped with plastic heels and they melt with alarming rapidity on BMWs.

The now obsolete silencer had a goodly cutaway underneath to allow removal of the rear axle and wheel in proper QD manner but the new one with its less pronounced dent is nowhere near as facilitating although BMW claim the QD action is still retained.

The frame is the already incorporated long wheelbase design and is a beaut. It does everything it is commanded to do, and even for a big bike the 900 can change direction with incredible agility. The front fork is equipped with an adjustable steering damper but one of the three R90S models I have sampled had a disturbing head shake — rather like the first Commandos had — when the throttle was rolled off and the hands were removed from the bar. Damping down the steering head turncock did not cure the problem, in fact selecting number two position on the dial gave another problem — it screwed the whole shooting match so tight it was like riding on automatic pilot. That was one bike in three though, and may well have been due to nothing more than an unusually unbalanced front wheel.

The ride given by the suspension is, if anything, a shade soft but now I'm splitting hairs and this error (if it is one) falls on the right side of the wall. Over bumpy roads — and the French have the biggest testing ground in the world — it was almost perfect, and although it never stepped out at the front end it did get a shade wallowy when the power was knocked off on fast tight bends when all the available road was used.

Medium-wrench braking resulted in the front end slamming down into the tarmac at an initially disturbing rate and leads me to believe that there is no real need for such soft springs inside such long legs. A better arrangement cannot be too hard to find but as the setup has no real vices — just an unusual characteristic — there may be no need for change. The alloy rims are shod with special high speed Metzelers capable of withstanding ton-up cruising for long periods of time. They have plenty of sidewall tread and can be laid over to the extreme, rocker box scratching the road, and still hold a firm line. Riding in heavy rain and even on black ice proved no hassle — although on the ice the power was kept evenly fed and the bike upright with no attempt at braking for bends. At 20mph the back end skittered and slid quite a bit but not once did it become uncontrollable. These boots are good.

Bearing in mind the comments about the engine's need for a fair amount of gearbox use, the power can be used two ways: controlled or brutally. Red line is at 7,000rpm and in all five gears the needle rockets around with absolute ease. Like a thoroughbred racer, the motor growls onto megga somewhere near 6,000rpm and that final thousand revs are attained almost as quickly as it takes you to search out the clutch lever and snick your boot under the gearshift pedal. Over-revving gets you nowhere at all and with such a collection of high-cost parts under your pants, t's as well to treat the bloodline as true max. Gearbox action, due to stiffer springs on the clutch and closer ratios on the cogs is still a fair way short of perfect but in BMW terms, is as near as you're going to get. Silent selections up and down are easily accomplished with the exception of second-to-first changes. If you are careful even these can be managed and should this aspect of the bike never be improved again,. I for one will be happy with what has arrived. Selection is positive and firm and has a nice feel to it. With the sole exception of the one "S" that wagged its head, neutral finding at rest and gear finding at speed were no problem. I mention the one sub-standard model again as this had a hard life from birth. Before it had a dozen miles under its wheels it was being abused up and down the Autoroute, thrashed by journalists, subjected to the vehicular equivalent of GBH in a series of five test days throughout the UK by prospective buyers or those come to worship (and destroy if at all possible), and fractures and bruises to its person informed one of at least one tumble's worth of damage on each side.

Secret of the uncrunched gearchange is, as previously, treading up or down to take up the spring-loaded play in the gear lever before making the change and then hooking it in fairly slowly as the clutch lever is pulled in.

The gearbox is the biggest new component in the machine and as such has to be proved despite the amount of development that may have gone into it.

TECHNICAL DATA

ENGINE	R 90/S
Output	65 bhp (DIN) at 7000 rpm 75 bhp (SAE) at 7000 rpm
Max torque	7.6 mkp at 5500 rpm
Bore	3,54"
Stroke	2,78"
Capacity	54,8 cu.in
Compression ratio	9.5:1
Carburettors	Dell'Orto PHM
TRANSMISSION	
Gearbox ratio I	4.40:1
II	2.86:1
III	2.07:1
IV	1.67:1
V	1.50:1
Final drive ratio	3.00:1
Tyres front	3.25 H 19
rear	4.00 H 18
Rims front	1.85 B x 19
rear	2.15 B x 18
BRAKES	
Front:	Twin disc brake with self centring caliper
Diameter	10,4"
Rear:	Full width light alloy hubs—simplex
Diameter	7,87"
Width	1,18"
Friction area	16,55 sq,in
DIMENSIONS	
Length	85,9"
Width	29,1"
Height	47,6"
Seat height	32,3"
Wheelbase	57,7"
WEIGHT	
(In road trim, full tank)	474lb
ELECTRICAL EQUIPMENT	
System	12 Volt
Alternator	280 Watt
Fuel tank capacity	5,28 imp.gal
Inc reserve	0,55 imp. gal
Optional	
Top speed	122/125mph

For a bike that really knocks on, the brakes are perfect. The twin 10¼in discs on the front hub are all you will ever need in any event, being light, progressive and immensely powerful. They are not prone to either locking or fading, the latter can only be detected after many miles of downhill racing or after eight full power stops one after the other. A real lockup was never achieved but tyre squealing at speeds over the ton was easily accomplished with the front stoppers. The rear brake was extremely good too and both discs and drum have an inspection window for checking pad and lining wear, but I have some misgivings about holes in hubs. Where there's a hole there's a route open for water and grit to enter. Maybe BMW have proved to themselves that the centrifugal forces and air ducting in the area of a fast spinning hub keep all the muck away from the holes, and if so then my thoughts are groundless but the mental picture I have of a carborundum mix of water and grit sandwiched between a brake lining and its braking surface gives me the shudders.

Fuel consumption is good for a near-litre sports motorcycle, the factory claiming 56mpg at a constant 68mph and us achieving about 50/52mpg at 70 (not under laboratory conditions you understand, just by filling up and reading the odometer). Average mpg from a normal 200 odd mile ride which included rather more than 50 per cent of it at high speed was 45/47. The approximations — for which we apologise — are the result of translating litres into gallons and trying to level out the speed/mileage balance after being baulked on the road by slower traffic, but then MIRA conditions whilst being clinically accurate, are by no means average.

What else does one say about a thoroughly nice, fast and enjoyable motorcycle? Giving a rundown on the major yet separate aspects of the R90S somehow detracts from the complete package. Good power needs qualifying as does good braking and steering. One can have all of these and yet they can be mis-matched because one of these can be too good for the rest of them. With the BM, the whole thing works well and feels right because every functional component has been matched to its counterpart. So well in fact, that it proves hard to summarise the bike. When I find the perfect motorcycle I'm sure it will feel quite ordinary because that's just how a motorcycle should feel.

And the R90S in that context is one of the most ordinary bikes I have yet enjoyed. m

Capsule Test

BMW R90S

How a sedate touring machine turns into a superbike with a super price.

At a time when everything costs more and nothing costs less we expect new motorcycle prices to rise. But an over thirty-four hundred dollar production machine? That's almost inconceivable. Yet, here it is: the BMW R90S—made of steel, priced like gold. The BMW Sport has been designed and built to extraordinary standards of quality and therefore may be worth every penny its makers are asking.

There's absolutely no doubt that the R90S is a remarkable motorcycle. It does, or can be made to do, almost everything as well and possibly better than any other road machine you can buy. The Sport is exceptional in numerous areas: speed, acceleration, comfort, cleanliness, operating economy, smoothness, quietness, braking, reliability and convenience.

The R90S delivers tremendous performance in a most pleasant manner. There isn't one motion or operation performed by the Sport that is in any way harsh. The engine develops enough honest horsepower to get the bike moving in excess of 120 mph, yet feels very mild. Veteran BMW riders recognize the engine's character but are astonished at the acceleration rate it produces. Those unfamiliar with BMW's flat-twin sense luxury but not much speed until they see the speedometer needle winding around like the dial on a Toledo scale. It's a sleeper.

As with any BMW, the Sport stands out in a crowd. The all-aluminum engine has its cylinders jutting like finned wings into an unobstructed airstream. The cylinder head and valve cover assembly is long (11 inches) and the total engine width is broad at 29 inches, about 5 inches wider than a Kawasaki Z-1. The unitized design of BMW's engine keeps all auxiliary components in one weatherproof housing. The only exposed components are the carburetors. BMW's unique crankcase hump hides and protects the air cleaner while eliminating intake drone and other normal running noises that contribute to the high sound levels associated with most big road machines.

The Sport is by far the strongest production motorcycle BMW has ever made. Given a handful of throttle, the Dell'Ortos inject fuel into the engine to balance the inrushing air and the Sport responds smartly, accelerates away strongly without a trace of stumbling. The healthiest part of the power range is from 5000 to 7500 rpm. Unlike most other Superbikes the R90S is free of harsh engine reaction when the trigger is pulled. There's no wailing intake drone, sharp exhaust growl, handgrips vibrating out to the size of baseball bats or trembling foot pegs. The R90S is to other Superbikes what a Boeing 747 is to an unlimited-class racing plane. The power is there, but only the muted sensation of speed is revealed.

The Sport frame is the same as last year's 750 with the two-inch longer swing arm. The R90S chassis has some additional gussets, but the geometry is unchanged. BMW uses large 1-5/16-inch OD tubing for most of the frame. Ugly brackets and tabs are almost non-existent, because the engine's crankcase hump contains most auxiliary components. The chassis design offers little rigidity from triangulation; it relies on large tubing for strength.

Comfort is supplied in abundance. There isn't another road machine we have ridden that matches, much less exceeds, the ride and seating comfort of the R90S. Though it is not as vibration-free as some, its riding position, handlebar shape and saddle design have been very well executed; they're not just afterthoughts. The attractive bikini fairing actually makes a larger contribution to comfort than to appearance. Un-faired machines don't offer any upper torso protection from wind buffeting. The R90S fairing eliminates all the buffeting and most cross-wind abuse absorbed from the waist to the neck. We were both surprised and pleased that the fairing worked so well. By deflecting the airflow off the rider's midsection, longer sustained high-speed (above 70 mph) riding time is possible without rest stops.

The R90S isn't perfect. The Sport produces a gentle buzz at its foot pegs and the handlebars shudder slightly. Engine vibration is still below the levels of the Z-1, 850 Moto Guzzi or any Harley V-twin.

Everyone notices the powerful double-disc front brake that looks, and is, capable of stopping the Sport from 60 mph in just over 130 feet (making the Sport the fifth-hardest stopper ever tested by *Cycle*). But few people notice that there's no master cylinder on the handlebar. BMW has placed it under the gas tank, just behind the head stock.

Very vew motorcycles like the BMW R90S exist today. A lot more could if manufacturers were pressed, or enthused enough about motorcycling, to develop sport-type machines. But have we arrived at the day of the $3400 production motorcycle? We have to wonder. ◉

BMW R90/6

All The Little Things That Have Bothered Us In The Past Have Disappeared. Does That Make The New BMW A Perfect Motorcycle?

Cycle World Road Test

IN THE PAST few years, there seems to have been a leaning toward the old saying "the bigger the better." While this logic doesn't always work out, it is infallible when applied to BMW's latest offering, the R90/6. Here's a machine that is no larger physically than the other machines in the BMW line-up, stops better than any of its predecessors, and goes like blazes as well.

The new BMW R90/6 is so exciting it's difficult to find a point at which to begin describing it. The horizontally opposed engine has been with BMW since its first model, built in 1923, but has gone through many changes since that time. The latest /6 series machines share many of the features of the recent /5 models, including the same basic crankcase casting and the same stroke; only the cylinder bore has been increased to 90mm to give an actual displacement of 898cc for the R90/6. The 70.6mm stroke is the same on all BMW motorcycles, from the 600cc R60/6 to the R90/6.

To make more room for the 90mm pistons, the crankshaft flyweights have a smaller outside diameter and tungsten plugs are used to obtain the proper balancing figure. And as the R75/5 first felt to us, it seems as though the size and/or weight of the main flywheel has been reduced. This freer revving and decelerating of the engine is most likely caused by the heavier and larger diameter pistons, however.

BMW still maintains its tradition of mounting all the major turning surfaces inside the engine in plain bearings, which are three-layer affairs composed of bronze, lead and indium. Plain bearings generate more heat than ball or roller bearings of equivalent dimensions, which means that there must be a copious supply of oil at hand to carry away the heat. As in the /5 series BMWs, an Eaton-type trochoid oil pump is mounted to the flywheel end of the camshaft. This pump is, in reality, a four-bladed impeller revolving inside a five-chambered housing, capable of moving vast quantities of oil at high pressures. At 6000 rpm the oil pump on the R90/6 is capable of circulating 370 gallons of oil per hour!

As on earlier models, the large-capacity oil filter is retained, featuring a full-flow pressure-release valve. This feature permits only filtered oil to pass back into the engine. Pressures from the crankcase are dealt with using an anti-turbulence chamber which leads to a diaphragm in the inlet manifold. This not

only helps reduce the BMWs emissions, but also helps keep oil from being forced passed the crankcase mating surfaces, dirtying the engine.

The only non-plain bearings in the engine itself are those used at the rocker arms: these are needle bearings, the value of which is, for a touring engine, questionable to us.

Besides increased physical size, and the changing of valve diameters, the only significant change to the cylinder heads has been to locate the carburetor mounting spigots some 6mm closer to the inlet valve. This moves the shin-banging Bing constant velocity carburetors that much farther away from the rider's legs.

Inspection of the engine's front cover, behind which are located the alternator, contact breaker points and some electrical equipment, will reveal five cast-in ribs and six slots. The ribs are placed there to help deaden any engine mechanical noises and the slots allow cooling air for the new 280-watt alternator to circulate through. Aside from the cover, the engine looks almost exactly like earlier models.

BMWs have had a reputation for noisy gear changes ever since we can remember. Well, that's a reputation that has been destroyed with the new, five-speed transmission, a unit now on par with many of the slick shifting Japanese gearboxes. Both transmission shafts are supported on enormous ball bearings and a newly designed gear selector quadrant slides the gears smoothly into engagement. First, 2nd and 5th gears are located on the mainshaft while 3rd and 4th gears ride on the layshaft.

Gone is the dreaded "clunk" during gear shifts: only a pleasant "snick, snick" is now heard, with the exception of a slight "snock" when down-shifting from 3rd to 2nd and 2nd to 1st. No more will motorists look over at you when you're shifting, expecting to see your transmission lying out in the middle of the road!

The dry, single-plate automotive-type clutch and diaphragm spring are retained, and even aside from design considerations, it's easy to see why. Ultra-smooth starts may be made and complete disengagement during gear shifts is assured. Clutch lever pressure is slightly higher than before, but will go unnoticed by most. Final drive remains unchanged. The driveshaft runs in an oil bath within the right hand swinging arm member and a universal joint is employed at the transmission output shaft to compensate for the swinging arm's up and down movement. Slight changes in the length of the driveshaft are compensated for by curved, helical teeth in an internally splined coupling shaft. Aside from infrequent oil changes, the drive mechanism is virtually maintenance free and is much cleaner and quieter in operation than an exposed chain, used on most of today's motorcycles.

The Bing carburetors are actually very precise in their operation, which depends on manifold vacuum, engine speed and a butterfly valve which controls the opening of a slide which is connected to a tapered needle. No matter how far the butterfly, which is the throttle valve, is opened, the slide and needle won't raise higher than to admit the maximum fuel/air mixture the engine can use, so flooding the engine by overambitious throttle twisting is impossible.

At the other end of the motorcycle are a pair of the largest and most efficient mufflers to be found on a two-wheeler. These have been slightly redesigned from the /5 series machines for silence with a minimum loss of power with the larger engine. Even though they emit a slightly more "nasal" sound, BMW claims that the total noise output of the R90/6 is still a very quiet 76dbA. Part of the pleasure of riding a BMW is cruising by traffic without disturbing anyone. Even your average "Upjohn Dewey" recognizes the fact that the BeeEm is one hell of a quiet machine.

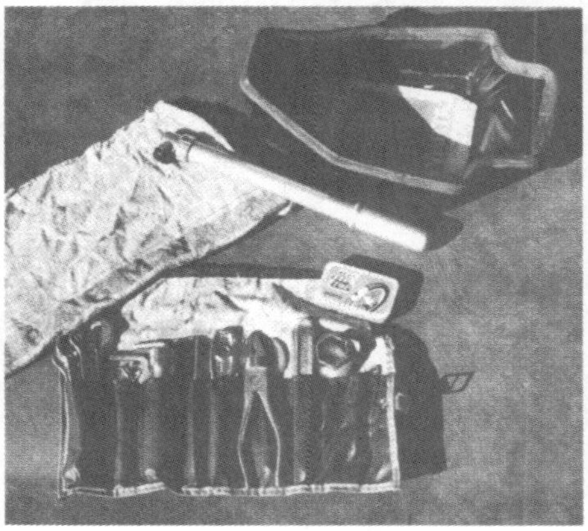

Another new item on the /6 BMWs is the alternator which has increased in output from 200 to 280 watts and a larger battery which has grown in power from 18 to 25 amp/hr. This unit is larger and heavier than the older batteries, but its increased power is well worth the weight. The electrics on the BMW are still among the finest in the world. A healthy 0.5 bhp (DIN) starter spins the engine over effortlessly, even on the coldest mornings, and a switch in the transmission is connected to a switch on the clutch lever to prevent the starter from being actuated if the motorcycle is in gear and the clutch lever isn't withdrawn.

Also improved is the headlight featuring the excellent beam patterns of before, but it's a tad larger at 7.09 in. in diameter.

You can't believe how well it lights the road on a dark night. And BMW has finally given us a separate speedometer and tachometer (time for celebrating!) with needles which don't flail wildly about while you're riding. Between the main units is a small dash panel full of vertically arranged idiot lights which warn of low brake fluid, low oil pressure, no alternator output, a neutral indicator light and a turn indicator light. The neutral light is not to be trusted, as it sometimes glows when the BeeEm is still in gear. BMW has finally done away with the their "standard" ignition "plug" (like the one everybody else with a BMW has) and has replaced it with a proper ignition key, inserted on the left hand side of the headlight shell. The same key locks both saddle and steering head for theft protection.

Underneath what is probably the most comfortable seat in the industry is a plastic trough containing a wonderous assortment of tools, quality tools besides. There's even enough extra room for a pair of gloves, a camera lens or what-have-you. On the inside of the top, left-hand frame rail is a hand tire pump; to our knowledge, the only mass produced motorcycle to be still so equipped.

It should be mentioned that the speedometer and tachometer are fitted with glare resistant lenses, a boon on sunny days, but nighttime illumination could be improved; they're a little on the dim side. Clutch and front brake levers and their supports are now finished in a matt, glare-resistant black instead of polished aluminum, another small touch.

If you'd asked a rider about the handling qualities of a BMW just a few years ago, you'd have probably gotten a smile and assorted wisecracks. But with the introduction of the /5 series BMWs in 1970, BMW started turning people's heads with their. vastly improved handling qualities. To underline that feeling, Reg Pridmore on the West Coast, and Kurt Liebmann on the East Coast, have been winning production road races on BMW R75/5 machines with amazing regularity. The /6 models have what is basically the same frame, but it is strengthened significantly in the steering head area. Gusset plates and long welded seams are still featured at critical areas, and yet the frame's weight is still down in the neighborhood of 30 lb. The massive swinging arm is still carried in adjustable, tapered roller bearings.

Suspension has always been a strong point with BMW. Even with the older Earles-type front fork, which provided better comfort than steering, the BMW has always been a luxury touring machine. The /5 and /6 models are no exception in the comfort department, except that the telescopic front forks, which provide almost 8 in. of well-damped travel, add immeasurably to the handling qualities, especially over rough roads. And the superior action of the 5 in. travel rear shock absorbers do their part in making the BMW one of the better handling big-bore street machines available today.

At lower speeds the rider is well aware of the fact that he is riding a large motorcycle; and that feeling pretty well remains

Front brake master cylinder is located beneath the fuel tank. Note substantial gussets on frame in the area of the steering head.

Engine front cover has five cast-in ribs to help deaden engine noise and six slots (three of which are visible) to allowing cooling air to circulate around the 280-watt alternator.

Improved gear selector mechanism helps obtain crisp, positive shifts every time.

Transmission contains a set of large, strong gears, and the main casting is smaller and lighter than older models.

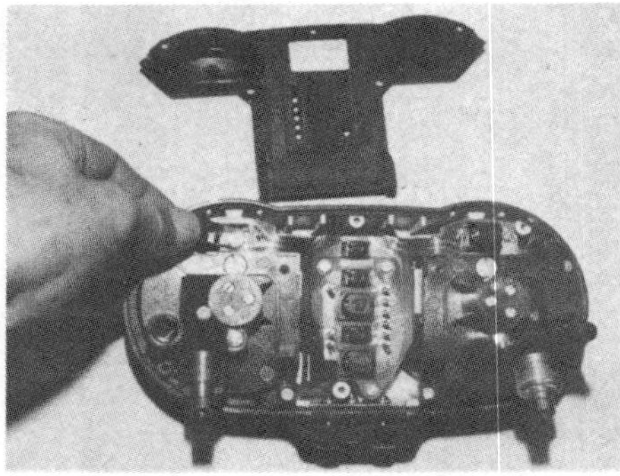

Back side of instrument panel shows easy to replace lamps and a printed circuit board instead of wires.

even when covering ground at a rapid rate. You don't "flick" a BeeEm from side to side through a series of "S" turns as you would a 350 Yamaha. Instead, the rider has to sort of force the BMW out of a straight line of flight. And yet, on the right road, the 900 can cover ground equally as fast as the smaller, more nimble machine. Add a few more variables such as potholes, ripples and lots of miles, and the BMW rider arrives at the destination sooner and far more relaxed and less tired than the fellow on the smaller bike. But the BeeEm doesn't have to pick on smaller machines; it'll raise havoc with the best of the Multis as well.

Previously using a friction-type steering damper on the /5 models, the /6 models feature a hydraulic steering damper with a three-way adjustment. But the new BMW steers so steadily that we almost felt the steering damper to be superfluous, except in tough crosswinds.

An area that BMW has previously not been overly strong in is the braking department. The brakes were adequate, but just barely for such a high speed machine. Now they've added a 10.4-in. diameter stainless steel brake disc and ATE caliper to the front wheel. The brake's master cylinder is out of sight under the fuel tank. Front brake actuation is provided by cable from the handlebar lever to the master cylinder, and hydraulic pressure takes over from the master cylinder to the brake caliper on the front fork leg.

Although the brake caliper unit is a single acting unit, it is permitted to "float" slightly in order to align itself with the disc, and it is mounted on an eccentric screw so that it may be adjusted to compensate for wear of the "dead" puck.

All motorcycles manufactured after Jan. 1, 1974 and sold in the United States for road use will have to have some method of checking the condition of the brake linings without removing the wheel. This is easy to do on a motorcycle equipped with disc brakes, but a conventional drum brake presents more of a problem. We've already seen some Japanese machines with indicator pointers attached to the brake cam pivot spindles, but this method isn't entirely foolproof because it is easy to change the indicator to a more favorable position although the brake lining is worn past limits.

BMW has solved this problem neatly by removing the chrome "hub cap" from the rear wheel. This permits the inspector to merely look inside and measure the thickness of the brake lining without removing the wheel. During the first few miles we rode the R90/6 the rear brake was a little insensitive, requiring a greater than usual pressure to make it operate properly. However, as the miles built up, so did the rear brake's effectiveness. Using the weight of the machine with a 160-lb. rider on board, (620 lb.), there is a very low brake loading of 5.5 lb. sq./in. Much fast riding in hilly country two-up was performed during our test and the brakes took the punishing treatment without a whimper. A machine that is capable of running as fast as the BMW R90/6 should be capable of stopping, relatively speaking, just as fast. This one does.

One amazing feature of the R90/6 is fuel economy.

In this day and age of gasoline shortage and possible rationing, it's nice to ride a vehicle that delivers 50 miles per gallon! Lower speed limits be damned. Take people out of their Eldorados and Continentals and put them on motorcycles and it'd help immeasurably.

Following recent tradition, BMW supplies its motorcycle with a ribbed tread pattern on the front tire and a more

BMW R90/6

SPECIFICATIONS

List price	$2950
Suspension, front	telescopic fork
Suspension, rear	swinging arm
Tire, front	3.25-19
Tire, rear	4.00-18
Brake, front, eff. dia. x width, in.	10.08 x 1.52
Brake, rear, dia. x width, in.	7.87 x 1.18
Total brake swept area, sq. in.	112
Brake loading, lb./sq. in. (160-lb. rider)	5.5
Engine, type	ohv opposed Twin
Bore x stroke, in., mm	3.54 x 2.77, 90 x 70.6
Piston displacement, cu. in., cc	54.8, 898
Compression ratio	9.0:1
Claimed bhp @ rpm	69 @ 6700
Claimed torque @ rpm, lb.-ft.	52.8 @ 5500
Carburetion	(2) 32mm Bing constant velocity
Ignition	coil and battery
Oil system	gear pump, wet sump
Oil capacity, pt.	4.75
Fuel capacity, U.S. gal.	4.74
Recommended fuel	premium
Starting system	electric, kick
Lighting system	12V alternator
Air filtration	dry treated paper
Clutch	single-plate, dry
Primary drive	none
Final drive	shaft and bevel gear
Gear ratios, overall:1	
5th	4.64
4th	5.16
3rd	6.40
2nd	8.84
1st	13.60
Wheelbase, in.	57.7
Seat height, in.	32.0
Seat width, in.	12.0
Handlebar width, in.	28.5
Footpeg height, in.	12.0
Ground clearance, in.	7.0
Curb weight (w/half-tank fuel), lb.	460
Weight bias, front/rear, percent	45/55
Test weight (fuel and rider), lb.	595
Mileage at completion of test	1150

TEST CONDITIONS

Air temperature, degrees F	62
Humidity, percent	41
Barometric pressure, in. hg.	29.99
Altitude above mean sea level, ft.	383
Wind velocity, mph	4-6
Strip alignment, relative wind:	

PERFORMANCE

Top speed (actual @ 7020 rpm), mph	115
Computed top speed in gears (@ 7200 rpm), mph:	
5th	118
4th	106
3rd	85
2nd	62
1st	40
Mph/1000 rpm, top gear	16.4
Engine revolutions/mile, top gear	3679
Piston speed (@ 7200 rpm), ft./min.	3325
Lb./hp (160-lb. rider)	8.99
Fuel consumption, mpg	50
Speedometer error:	
50 mph indicated, actually	46
60 mph indicated, actually	54
70 mph indicated, actually	64
Braking distance:	
from 30 mph, ft.	34
from 60 mph, ft.	132
Acceleration, zero to:	
30 mph, sec.	2.8
40 mph, sec.	3.8
50 mph, sec.	4.9
60 mph, sec.	5.8
70 mph, sec.	7.2
80 mph, sec.	9.0
90 mph, sec.	11.5
100 mph, sec.	14.9
Standing one-eighth mile, sec.	9.55
terminal speed, mph	83.63
Standing one-quarter mile, sec.	13.45
terminal speed, mph	97.27

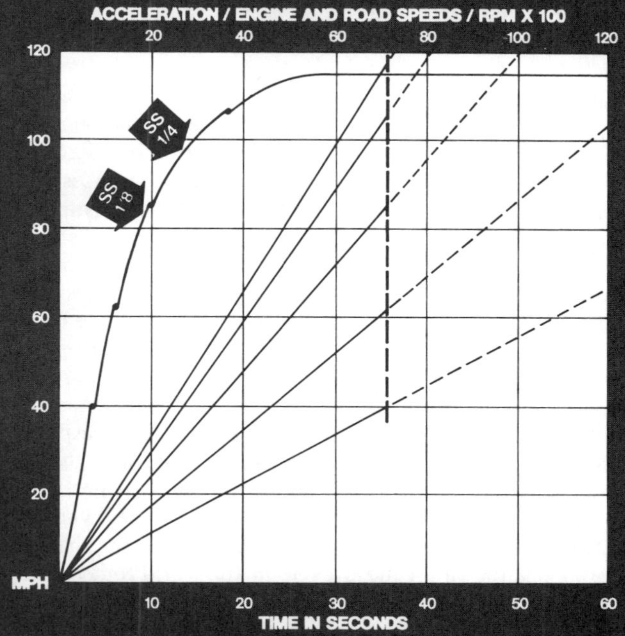

"blocked" tread pattern on the rear tire. The tires, incidentally, have an H rating, which means that they are capable of sustained running at speeds of 130 mph. Other tire designations include the R rating, good for sustained speeds of 90 mph; the S rating, for speeds of 110 mph; and the V rating, which is for racing tires.

The Germans have rarely done anything halfway, preferring instead to produce a complete, *final* motorcycle before releasing it for sale. The R90/6 is no exception. The extra weight gained from the addition of a fifth speed in the gearbox and a larger battery has been almost completely negated by juducious trimming and pruning off ounces wherever possible. The transmission casting is a prime example. The casting is slightly different and slimmer with less material around certain of the bosses. No, there has been no loss of strength, just a little precious weight. And, of course, such items as the fiberglass fenders help reduce the overall weight of the bike, too. The BMW R90/6 is the lightest machine of its size and horsepower around.

Another strong point of the BMW is the immaculate detailing. There are no rough welds, no mismatched castings; paint is carefully and generously applied. The chrome plating is flawless and appears very substantial, and there are no items which seem to have been put in the wrong place. The owner of a /6 series BMW will have a sense of pride in knowing that he owns probably the finest motorcycle in the world today. And with the myriad of motorcycles still being manufactured, that's saying a lot.

It should also be mentioned that the BMWs are available in several colors, but the large, 5.8-gal. fuel tank, preferred by many touring riders, will only be available at extra cost on the blue and black machines.

A look at the performance figures will show you that BMW is serious in its intentions of producing the lightest, best handling Superbike around. The R90/6 was one machine we really hated to return, and we look forward to testing the sport model, called the R90S, when one becomes available.

PARTS PRICING

Warranty	6 mo./6000 miles
Major Tuneup	$28.50
Air Filter Element	$7.50
Rear Tire (standard)	$33.38
Headlight Bulb	N.A.
Taillight Bulb	N.A.
Turn Indicator Bulbs	50 cents
Battery	N.A.
Clutch Cable	$5.55
Throttle Cables (each)	$6
Brake Cables	$5.45
Ignition Parts	
Points	$3.95
Condenser	$3.10
High Tension Coils (each)	$19.85

LONG DISTANCE TEST

BMW R75/5

Alastair Noyle mortgaged his house to buy a BMW. Was the gamble wise? Read on . . .

PHOTOGRAPHY DEREK SODEN

IT AIN'T easy — you try and assess your own precious bike and be honest! Living with a BMW can be worse than most; it'll tie you up in a welter of mixed feelings none of which can be seen clearly for the figures on your bank statement.

Believe it or not, a whole year ago when I bought the R75/5, it had come down in price due to VAT replacing purchase tax, and what is more, it has stayed there ever since. It was the reduction that decided the issue, we took the plunge and mortgaged the house.

If road tests are proportional to the amount of persuasion necessary to part you from your money, then the Bee-Em just had to be my choice; the R/5 series were at that time about the most written-up machines on offer, and all the comment was favourable. The message was loud and clear — quality, and all it implies, from sound engineering to that elusive trendiness of the Sunday supplement advertising.

The machine exudes refinement from every functional line, providing you remove those gimmicky chrome side panels. In times when the number of cylinders tends to be the measure of a bike, the apparent lack of metal raises the first doubt regarding value for money, for the machine is about as uncluttered as it is possible to make any prestige motorcycle.

(Top) An Avon fairing helped jack the price of Alastair Noyle's R75/5 to £1,255, but gives plenty of comfort and weather protection. He also fitted high American bars. (Centre) Tool kit fits easily in large under-seat tray. And there's even a tyre pump.

(Above) Standard headlamp, which gave a meagre beam, was replaced with this Cibie Biode unit to gain enough candle-power for fast night cruising.

A year ago last March saw my wife and me in London to exchange our Yamaha 650 for the R75/5. We got a good part-ex offer that made the journey from South Wales worthwhile. The trip back was one of sharp contrasts and the first wasn't good; a non-existent front brake plus cross-wired indicators and the North Circular Road at Saturday lunch-time. It all conspired to produce a traumatic introduction to the BMW experience.

I was getting a distinctly furrowed brow by the time we reached the second set of traffic lights. The A406 was wet for the first time in a fortnight, and I needed 300 yards in which to stop the heavily laden machine! This may seem exaggerated, but after the XS-2 front disc, it all seemed a little hairy. Fortunately the rear brake had more bite in it. Nobody mentioned to me that BMWs have virtually no brakes for the first hundred miles or so, a thing of the past now that discs are "in".

Besides specifying a black finish and a large tank, we had ordered a white Avon fairing. Fitting an Avon to a Bee-Em is no easy task, and I suspect the job was rushed at the last minute — hence the electrical problems. Another side effect was apparent to me as a newcomer to fairings, and this was the impression that the steering was rigid because I couldn't see the front wheel; daft maybe, but possibly enhanced by the short flat handlebar which isn't my favourite, particularly two-up.

I should probably have checked the machine more carefully before leaving, but isn't it reasonable to expect that a bike totalling £1255 be perfect on delivery? Perhaps the fact that the pre-delivery check card is still in the handbook was something to do with it.

We eventually negotiated the North Circular and set off west down the M4 — where we almost ran out of petrol. With some relief I switched to the ample reserve; we had spent the upward journey wondering if the Yamaha would stretch the distance between services with its ridiculously small tank.

The journey from Heston to the Severn was great, this contrast being good, very good. Most memorable was the quiet, vibration-free way the machine consumed the miles, even at running-in speeds — 4000 rpm matching 65 mph on the clock.

It wasn't until we crossed the Severn Bridge after dark that we learned of the final sad contrast with the Yam — the headlight was diabolical! Later, I discovered yet more bird's nest wiring.

I haven't mentioned the gearbox and its attendant clunk. I had been conditioned to expect it, and wasn't disappointed. Quite honestly, I regarded the noise almost affectionately, for it seemed like an anchor by which I could hang on to the BMW mystique in the midst of strange contradictions.

Once home the indicators were quickly uncrossed, though the headlamp took a little longer. It wasn't until replacing the switches in order to fit an "American" handlebar that I finally sorted the mess out; somehow the high and dip filaments were lit simultaneously at half power. Things were better afterwards, but not much! Only a Cibie Biode cured that. Now the illumination can only be described as fantastic. Within 300 miles the front brake had bedded-in to become the superb stopper it is now, provided you are careful on damp mornings — just one of the BMW foibles that can awaken you rudely out of the breakfast reverie.

First service came at 600 miles. It was carried out by a comparatively local dealer to spare another motorway dash; BMW-UK are helpful in arranging such things.

Shortly after the service a tappet began to make itself audible, embarrassing in such a quiet machine. A general check-over was indicated, and a chance to use the splendid tool-kit and handbook — would that the Japanese produced such tomes as this last item.

I commenced my check with the necessary task of fitting a mud-flap on the rear guard, and removed the wheel to make the job easy. Slipping that wheel off was a blissful exercise. Then I was unpleasantly surprised to find the interior of the brake drum plastered with oil. Hypoy 90 was seeping from the shaft housing into the bevel-drive housing and the external bleed couldn't cope with it. There were at least 60 cc's too much oil in the bevel. "Oil seal!" I thought, no little agitated at the prospect of my beautiful BMW in need of treatment so early in life.

To cut a long story and numerous letters short, I have since been informed that the fault is due to over-filling and extreme pressure. At the 4000-mile service my dealer expressed thinly veiled scepticism — boredom even — at the mention of oil seals. Yes, seals — my gearbox was weeping in sympathy by now.

"You've been flogging it," said the mechanic after reprimanding me for turning up unannounced. I had carelessly left a letter from him arranging the service and discussing the problem, 200 miles down the M4.

"We'll top the oil up to the correct level — our level!" I was too cowed by now to argue, as if I hadn't tried that. It all rather negated my efforts to get something done within the guarantee period.

The overall service charge came to an awful £16.16, which included a new contact breaker set at a more reasonable £2.45.

The last few months have changed my awe-stricken attitude toward the BMW charisma. Living with a machine that promises to be expensive to repair can do that, particularly if you have been conditioned to believe the lovely creature to be flawless. It is a melancholy thought that prior to this I had three Japanese bikes, none of which shed a single drop of oil.

My latest move against the dribbles entailed removing the hub myself. I took it to Bridgend where the lads at TWS were more than helpful. With the correct tool, they torqued-up the oil seal retaining ring by a whole turn. They wanted to change the seal, but I was curious, and wanted to see what would happen next by elimination. Well, I can tell you what happens next — new oil seals!

The R75/5 is certainly a machine that demands to be ridden — funny how some dealers suggest you have been flogging it, should flaws show up. It is now clear that the manufacturers wish to emphasise the sporting nature of the /5 series onwards, and live down the touring image of yore. Fair enough, but it is a pity that they prohibit the use of a sidecar.

Forethought in design is apparent at the most prosaic level, at the moment one approaches with bucket and cloth. The current BMWs must rank among the most easily cleaned bikes ever. Visually it is a crowd puller almost anywhere you park, and non-motorcyclists are as attracted as the afficionado.

The Avon fairing lends an expansive aspect to the otherwise trim frame, and may account for a slight low speed heaviness. It also makes swinging the machine onto the centre-stand more of a chore, though it is a knack worth cultivating as the fly-away prop-stand can't be trusted.

On the road, whether running solo or pillion, and providing the machine isn't hard pushed, the handling is very reasonable. Torque reaction, though much talked of, is negligible at all times.

The riding position at first seemed perfect for me with the high American-style handlebar, but as I began to press the R75 harder, handling faults became unpleasantly apparent. A mild tendency to understeer became pronounced and the rear suspension aggravated the situation with an unnerving sponginess.

I fitted an Avon SM rear tyre at 7,500 miles. The Continental had given excellent service, but I was unable to obtain one locally. The new boot introduced another horror — a high speed weave that insisted upon becoming a full-blown tankslapper when I tried to accelerate through it. I was riding solo in a group at the time . . . I've never seen a road empty so fast.

On the old tyre only the previous week we had been able to cruise effortlessly at 110 mph. I have since fitted a Dunlop TT100 as a Continental was still unavailable, and I think an hydraulic steering damper would be a good idea. No wonder the Americans demanded a longer wheelbase, though I believe Munich's advice is best: don't muck about with tyres!

Happily only one other fault has materialised to date; the battery developed an excessive thirst for distilled water which indicated voltage regulator trouble. I found the points solid with ash. The charging light still flickers occasionally to let me know possible trouble is lurking.

Although I still haven't entirely mastered the clunks in the gearbox, things are improving. As far as this model is concerned, the ratios are not perfect, low being too far from second on downward changes. A rapid trip down through the box can result in neck jarring deceleration, thanks to the very positive shaft drive.

Surprisingly, top is too high to allow the kind of 20 mph dawdling that may seem natural to a machine of this capacity. The motor is quite rev-happy, easily rocketing up to within 500 rpm of the 6750 red-line to give 40 in first. Changing up before about 35 mph can result in a marked drop in power, especially with a passenger.

The clutch is always positive, though rather more so after standing for a period. In the best tradition of dry-plate clutches there is no stiction early morning, and it never becomes hot-tempered in traffic.

The rear brake is excellent; it is progressive and has more feel than I have experienced before — very useful in the wet. Fuel consumption is remarkably stable at 50-52 mpg in all going, and oil is used in miniscule amounts.

The R75/5 has covered a varied 10,000 miles since March '73. Has the superlative machine fulfilled its promise? I know I can set out on a journey and expect to arrive relatively fresh, clean and dry. But beside being an efficient foot warmer, is the BMW value for money?

A year ago the question might have put me in a very awkward spot, but now that all the erstwhile "cheap" superbikes are not far from the magic grand in price, the point on this /5 model is debatable — the Kawasaki Z1 is currently £49 more expensive. If you are in the market for a large capacity machine, the /5 is still available and must be considered attractive.

What of the new /6 models? Well, my major complaints have been eliminated: both lights and generator are superior, so most definitely is the gearbox, but the price is superior too. If you have the yen to own one you have good taste, but second mortgages are harder to come by this year.

I am almost too close to be objective about the value. One or two disappointments have forced me to wonder, then I draw back from the awful admission — perhaps the legend that goes with the badge leads you to expect too much. Let's just say that I have no desire to be parted from my R75/5, it is that kind of well-bred machine.

Checkout

Engine	**Four-stroke, horizontally opposed twin**
Bore & stroke	**82 x 70.6 mm**
Capacity	**745 cc**
Comp. ratio	**9.0:1**
Carburettors	**2 x Bing constant velocity**
BHP at RPM	**57 at 6400**
Primary drive	**Direct via flywheel and single plate dry clutch**
Final drive	**Shaft enclosed in right swing arm tube. Universal joint at gearbox end and hypoid gear coupling at drive end**
Transmission ratios	**(1) 3.896:1 (2) 2.578:1 (3) 1.875:1 (4) 1.5:1**
Rear drive ratio	**1:3.2**
Electrics	**Alternator. Battery — Varta 12v. 15Ah**
Lighting	**Bosch (standard) Cibie Biode (fitted)**
DIMENSIONS	
Overall length	**82.7 ins**
Wheelbase	**54.5 ins**
Seat height	**33.0 ins**
Ground clearance	**8.0 ins**
Peg to seat	**21.0 ins**
Handlebar height	**38.0 ins**
Weight (full tank)	**506 lbs**
Oil capacity	**4.7 pints**
Fuel capacity	**6.35 gallons**
CYCLE PARTS	
Wheels (front)	**3.25 x 19 Continental tyre**
(rear)	**4.00 x 18 Continental tyre (standard) 4.25 x 18 Dunlop TT 100 (fitted)**
Brakes (front)	**Two leading shoe 7.9 ins**
(rear)	**Single leading shoe 7.9 ins**
EQUIPMENT	
Steering damper	**yes, friction**
Trafficators	**yes**
Electric starter	**yes**
Trip odometer	**no**
Steering lock	**yes**
Helmet lock	**no**
Headlamp flasher	**yes**
Kill button	**no**
Tool kit	**yes**
Other:	**Avon fairing, fly-off prop-stand, centre stand**
PERFORMANCE (Approximate figures)	
Top speed	**110 mph**
Standing ¼	**14.1 secs**
Fuel consumption (overall)	**52 mpg**
(driven hard)	**50 mpg**
Braking distance (from 30 mph)	**27 ft.**
(from 60 mph)	**125 ft.**
PRICE/WARRANTY	**£1,147 inc. VAT**
6 months parts and labour	

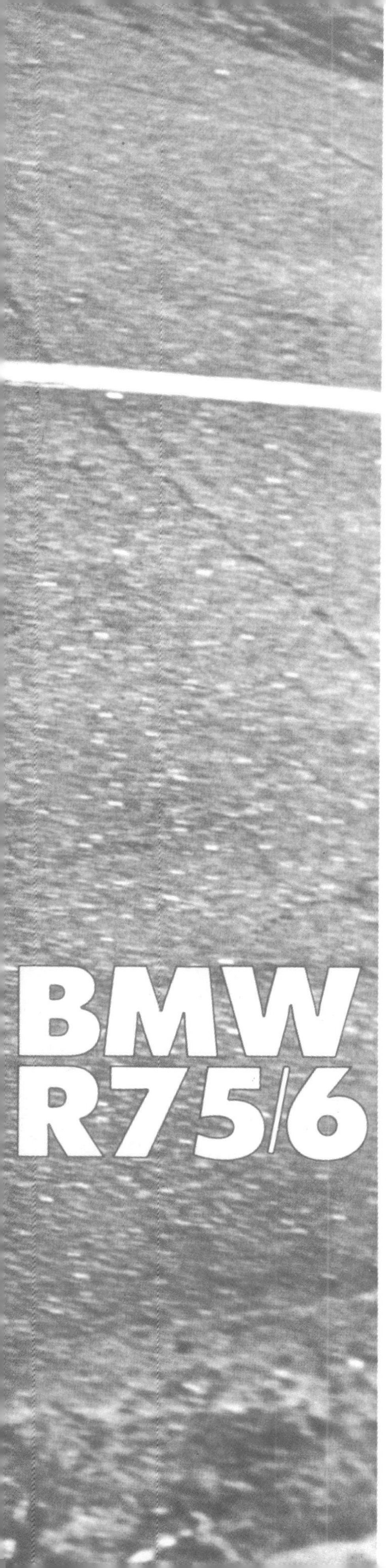

BMW R75/6

CYCLE RIDER **ROAD TEST**

FOR MANY YEARS the name BMW was synonymous with a very distinctive and different motorcycle. To many it appeared antiquated when compared with other contemporary two-wheelers. The bike was available in a choice of colors—black. Until this year the twin cylinder opposed engine made do with a four-speed gearbox, designed to rattle the nerves of all but the most devoted during gear changes. Internally expanding shoe brakes were standard at a time when Japan was putting discs on bikes as small as 200s. A unique Earles fork was used for the front suspension. Nonetheless, for the dyed in the wool devotee, the BMW was a truly beautiful creation. The only problem has been that there just weren't that many enthusiasts overwhelmed by the BMW design. It became obvious that something had to be done, both mechanically and aesthetically, to move the name BMW forward a peg or two. Over the past few years BMW, a company who is "set in their ways," has made changes. To some BMW enthusiasts these changes were an outrage.

The Earles fork was replaced by a telescopic unit; other changes were made to improve the appearance and handling qualities of this chassis. It is possible to purchase the BMW in seven different colors, instead of the traditional black. Is nothing sacred? Evidently not, when upgrading the product is the motive. The cumulative efforts of the last several years are reflected in the 1974 model line. To celebrate the 50th anniversary of their horizontally opposed twin, BMW extended invitations to several of the major motorcycle publications including *Cycle Rider* to fly to Germany to attend the unveiling of their '74 models. On the flight over speculation ranged from a four-cylinder opposed unit to a rotary shaftie. By now everyone knows what was waiting for us. The two new 900s weren't as exotic as we had anticipated but BMW had come up with four attractive updated tourers.

The painted surface of the BMW is truly immaculate. All seven colors sparkle with absolutely no orange peel effect or flaws in the surface. The pin striping around the tank and fenders is done by hand, an example of BMW's old world craftsmanship. The chrome is also top quality. Unfortunately the chromed head pipes do tend to discolor. To disguise this somewhat, a cheap looking heat shield is attached to the pipe. This appears to be a last minute effort.

The BM's have been the quietest bikes around for years. To make them even quieter they have utilized a crossover tube between the two headpipes. This also tends to reduce engine vibration, which is almost nonexistent anyway.

The most annoying noise that was encountered during the 2,000 mile test emanated from the engine—the tappets to be more specific. We felt certain that this loud "British twin" type noise was not intended as standard equipment. The clearance between the tappets and the valves was readjusted and the noise disappeared.

The engine design launches BMW into its second half century of involvement with the horizontally opposed, air cooled, two-cylinder powerplant. The 750 engine has an enormous bore of 82mm and a stroke of 70.6mm. For a supposedly mild mannered touring

Some people feel that BMW stands for Best Motorcycle in the World. They may have a point there.

machine the BMW does sport a rather high compression ratio of 9.0:1. Claimed engine output is 57 horsepower at 6400 rpms. Engine redline at 7,000 is indicated on the tachometer. The engine which looks on the ponderous side is actually quite light. The aluminum shrouding conceals the alternator and points on the front of the engine between the downtubes.

The air cleaner is mounted just below the gas tank. It feeds two 32mm Bing constant velocity carburetors. Surprisingly, these carbs worked without hesitation or any of the other ills that are sometimes inherent to a constant velocity carb.

An electric starter is new this year. We found that for cold starts it was necessary to operate the choke to its maximum. It is located just above the intake tract on the left side of the aluminum engine case. Leaving the throttle closed and depressing the starter button fires up the engine, which will idle happily without any assistance. It was necessary to slowly rotate the choke lever to the off position as the engine warmed up, otherwise the engine tended to run extremely fast. We found it virtually impossible to start the BMW while cold with the kickstarter. The starter lever is located on the left side of the engine and must be kicked to the outside, rather than to the rear, as with most machines. It seemed the kickstarter ratio was not great enough to turn the engine over at an adequate rate of speed. Thank goodness for electric starters.

A somewhat unique feature that BMW has incorporated is a lockout that prevents the engine from being started unless the gearbox is in neutral. Otherwise nothing happens when the starter button is pressed.

In the past the BMW has been known to have a rather severe acting clutch. This criticism should go down the drain with the '74 models. We found that engagement was not as hard as the previous models, and that the amount of pressure required to operate the clutch lever was about average. Unlike the majority of motorcycles today that utilize a wet multi-plate type of clutch, BMW's clutch is a dry single disc unit similar to automotive clutches.

BMW finally went to a five-speed gear box this year. All we can say is: it's about time. The shifting action of the four-speed, to say the least, was unrealistic. It wasn't all that difficult, but it was accompanied with loud clunking noises. We get letters from readers claiming that they can shift the four-speed unit without clunking, but we can't number ourselves among this elite group.

First gear can now be selected without any effort. In fact the only

way we knew that the gearbox was engaged was when the neutral light went out. A rider that is insensitive to the mechanical operation of the machine can still clunk and grind each gear change if he's not careful. After the first day with our test machine we were able to shift from first to fifth and back again without the customary clunk. All it takes is a little timing on the rider's part, and a familiarization with the machine.

Neutral is extremely easy to find. At a standstill neutral is automatically selected when shifting from first toward second gear. It didn't matter how much effort and speed are put into this shift.

BMW is a staunch believer in shaft drive rather than the normal chain drive. The designers and key personnel at BMW are definitely opposed to using chain drive. Both systems offer their advantages and disadvantages.

BMW R75/6

The shaft drive system does run quietly. First time riders are usually quite surprised upon acceleration to find that the whole motorcycle feels like it is lifting straight up to the top of its suspension travel. This is created by the driving gear trying to climb above the driven gear in the rear end. Once the initial surprise subsides it no longer startles, and it does not detract from the on the road performance of the BMW.

We are glad to report that the R75/6 did not leak one drop of oil during the entirety of the 2,000-mile test. We did get some seepage around several points, but it was not excessive. Nothing more than collecting dust.

Under the 3,000 rpm mark we noticed some shake from the engine, similar to that of a rubber mounted engine. This was never any trouble since 3,000 rpm is the bare minimum at which this engine should be operated.

The engine is held in place by a tubular double cradle frame that is surprisingly stable. Perhaps one factor contributing to this stability is the swinging arm. It is supported by tapered bearings which make it possible to completely remove all side play. Of course the swing arm is just one of the several things that make the BMW handle well. Every single frame that is assembled at the factory is checked for alignment. If it does not meet certain specifications it is returned to be straightened. The material is a high grade of steel which also reduces any flexing.

Some manufacturers seem to believe it is necessary to have a rigidly suspended machine for good handling. BMW doesn't. The front forks, complete with rubber gaiters, allow an enormous eight inches of travel. Not only is this travel "excessive" compared to other machines, but the action is soft.

At the rear end we find three-way adjustable shocks with considerable travel, five inches. Many may find it hard to believe that we found the R75/6 one of the best handling motorcycles we have ridden. It is by far the most comfortable.

Another major and long overdue change for the BMW is the addition of a ten-inch hydraulic disc brake up front. Here again BMW has taken somewhat of a different approach compared to the majority of units on the road today. Instead of mounting the heavy master cylinder in front of the forks they have positioned it to the rear. This helps move weight further back in the interest of improving low speed handling. The disc has somewhat of a spongy feel when compared to the brakes fitted to Japanese bikes. We imagine that this is attributable to the fact that the brake fluid reservoir is not connected directly to the brake lever. Instead, it is operated by the brake cable. The reservoir is positioned on the frame beneath the gas tank. This placement does do away with the unsightly reservoir, which is generally mounted to the handlebar, and also protects it in the event of a fall.

The rear brakes are still a single leading shoe arrangement. They work quite well. They rate among the better brakes that we have yet to test. To many the left side of the wheel hub will look unfinished. Instead of fitting a brake wear indicator to the brake hub itself, BMW has cut inspection windows into the wheel so brake wear can be checked by inspection of the linings.

The most obvious drawback to having the brake reservoir hidden is that it is difficult to check fluid level. In an effort to ease this chore BMW has fitted a light to their redesigned instrument panel to indicate when and if the brake fluid level drops below the recommended minimum.

The instruments are now in a cluster instead of being mounted in the headlight, and include, in addition to the tachometer and speedometer, several warning lights, ignition, neutral, brake fluid level, high beam and flashers. During the daytime all of these instruments are extremely easy to read. The glass that covers each is slightly frosted to prevent any glare from the sun being reflected into the rider's eyes. The well lighted speedometer is easy to read at night, but the tachometer is rather dim.

The ignition switch is located on the left side of the headlight. Instead of having three positions (on, off, and park) it has a total of five; off, parking light, ignition, parking light with ignition, and headlights with ignition. The key, by the way, is a radical improvement over the large knob generally used for operating the ignition.

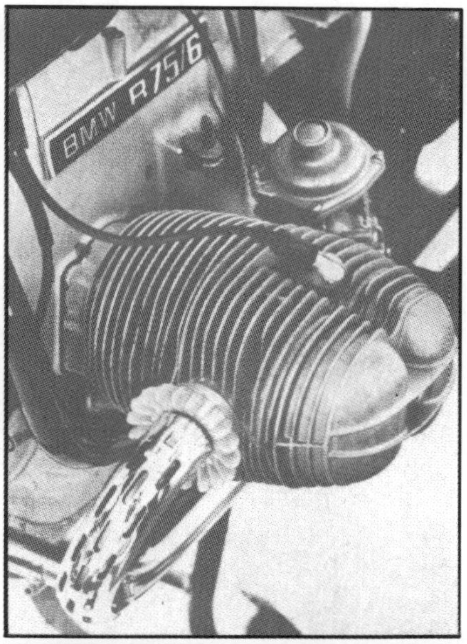

BMW R75/6

It now looks and feels like a key. The electric starter button is located on the right side of the handlebars along with the turn signal switch. These are incorporated in one unit. On the left handlebar is the dimmer switch and horn button, also in one unit. First gear can be selected with so little effort that it is impossible to detect engagement of the gearbox. As the clutch is slowly engaged and the bike begins to move slowly forward it is very apparent to the rider that the BMW is extremely light and agile at slow speed despite its appearance. Aside from the Suzuki T500, the BMW has perhaps the lightest feel up front, and is the easiest steering motorcycle at slow speeds that we have encountered. Perhaps the first couple of gear changes will clunk, but after that it is an easy matter to time each shift and prevent this annoyance.

If the throttle is opened quickly, particularly in the lower gears, the sensation of the machine jumping off the ground becomes quite apparent. Once in motion there is no tendency for the machine to shift from one side to the other when accelerating as some people who fear the torque reaction of the horizontally opposed engine may believe.

Our 150-pound testers preferred the rear suspension set in the bottom (softest) position. For two-up we went to the middle position. With a heavier rider the top position is necessary.

When running the BMW down any road at any speed one thing becomes quite apparent, the smoothness of the ride. The ultra long travel suspension units work as a team and tend to smooth out even the roughest of highways. We found that there is good damping control. This is essential with the long stroke in both the front and rear suspension systems.

The saddle has been modified somewhat this year to allow for a more comfortable seating position. It was possible to ride for many miles at a time without becoming fatigued.

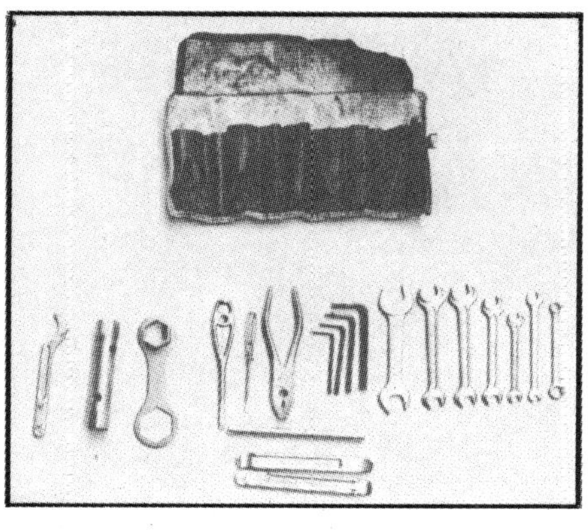

Because of the width of the cylinder and head it is not necessary to use folding footpegs. The BeeEm's are rigidly mounted. When sitting astride the bike in a normal riding position we had no problem as far as our shins coming into contact with the carburetor or the cylinder. We did find it possible to apply the rear brake and get our toe hung up on the exhaust pipe if we were trying to keep our foot in too close to the engine. This could be serious in some circumstances, but we adjusted our riding position accordingly and had no more trouble.

The double saddle allows a more than generous amount of room for two riders. Peg placement for the passenger is somewhat further forward than we would like to see. We found the passenger's toes under our heels. Because of this it was necessary for the passenger to ride with their toes on the pegs.

In looking over any motorcycle an individual can always find fault with something or another. The BMW is no exception. The only faults we can find aesthetically would be called nitpicking. Mechanically, the top end of the engine was somewhat noisy until the tappets were properly adjusted. The exhaust noise is almost nonexistant at cruising speed. The brakes were good. BMW has put together an extremely fine road machine that now offers both styling and performance. The average mileage for 2,000 miles was 43mpg. This is about average for touring motorcycles of this size at an indicated 60mph. With the almost complete absence of vibration, and the extremely smooth ride, the BMW makes by far the finest and smoothest touring machine that we have encountered to date. ■

BMW R75/6

CYCLE GUIDE ROAD TEST

BMW R90/6

BMW climbs on the big-bike bandwagon.

Bavarian Motor Works has been around for a long time making motorcycles, cars, and airplanes (The BMW logo is based on an airplane propeller.), among other things. For years their motorcycles have been considered the epitome of reliability, comfort, and elegance. But more recently, without sacrificing its reputation for civility, the West German marque has been cultivating a sporting image for its motorcycles.

One of the results of BMW's change in attitude is an increasing use of their machines as sporting motorcycles. More BMW cafe racers and production road racers are appearing on the scene all the time.

For the past few years BMW has been active in local and national championship road racing. Their record on the national circuit hasn't been spectacular, but their West Coast rider, Reg Pridmore, has dominated California open class production road racing to the point that when he's present with his 750 BMW, the only question is who will finish second. At the two production races at AMA nationals last year (Laguna Seca and Pocono), Pridmore brought the BMW home third behind factory Kawasaki riders.

Another result of BMW's sports orientation was the addition of two new 900cc models to the top of their line: the R90S, a cafe racer, and the subject of this test, the R90/6.

THE BIKE: The basic BMW layout—that is, the horizontally-opposed flat twin engine, utilizing a driveshaft—is currently unique to BMW in this country despite its use in the past by Marusho, Zundapp, and Ural, among others (although Honda has been experimenting with a shaft-drive flat four).

The engine-transmission unit looks huge and heavy, but its appearance is deceiving. Much of the room inside the cases is taken up by a large airbox, and the illusion of bulk is added to by the horizontal cylinders which protrude out into the airstream, thus keeping cool. The engine is only about 29 inches wide, but 11 inches on each side are taken up by the cylinder and head. Because the crankshaft is set parallel to the direction of travel and the crank throws are offset, the cylinders are offset—the left one is slightly ahead of the right.

The engine is basically a bored-out 600cc BMW. The 70.6mm stroke is common to the entire BMW line, and the R90/6 has a bore of 90mm. Pushrods operate overhead valves, and the 180-degree crankshaft keeps the pistons opposed, thereby counterbalancing each other.

Various accessories are located inside the engine cases. The alternator and points are mounted under the large cover at the front of the engine; the electric starter is inside the case above the engine; and the back-up non-primary kick starter is located on the left side. The starter lever swings outward and is difficult to use, but fortunately you don't need it very often.

Constant-velocity 32mm Bing carburetors are mounted at the rear of each cylinder. Two plastic hoses connect the carbs to the airbox. Two petcocks, one on each side of the tank, feed the carbs. The markings on the petcocks are "A" for on, "ZU" for off, and "R" for reserve. The

PHOTOGRAPHY BY MARC MADOW AND ART FRIEDMAN

choke lever is mounted on the left side of the engine case just under the gas tank.

The engine layout is in the BMW tradition, but major changes have been made elsewhere on the bike. A standard single-leading-shoe drum brake stops the rear wheel, but the single thin disc brake mounted on the left side of the front wheel is obviously different. This disc brake system differs from most others in two ways: (1) The master cylinder is located under the gas tank and connected to the lever by a cable. Its location protects it in the event of a crash, but more importantly, conceals it from vandals, who lately have been ripping off master-cylinder caps like they break automobile radio antennas. (A warning light in the instrument panel also warns you if the fluid level gets low.) (2) The caliper is mounted behind the fork leg. In theory, this minimizes distortion of the fork leg during braking and the consequent tendency to pull to one side. It also puts the weight of the caliper closer to the steering axis where it has less effect on steering action.

Another major change for BMW this year is the five-speed gearbox. Although few people ever complained about gaps in the ratios of BMW's four-speed, a five-speed better fits the new BMW image.

Quite a few styling changes have also been made. The chrome panels on the gas tanks of last year's bike are gone, replaced by paint and pinstripes (in seven available colors) to match the fiberglass fenders and side panels. The tank itself is tall, fairly narrow, and holds 4.8 gallons. (A 5.8 gallon tank is available as original equipment for an additional $40.) The tank is easily removed to service the front brake master cylinder.

Another stand-out styling change is the instrument panel, which is in its own case and uses printed circuitry in place of the internal wiring of the speedometer and tachometer. The printed circuits are more resistant to water and vibration. Between the speedometer and tach are five idiot lights: neutral, turn indicator, oil pressure, generator, and brake failure. Also new is a real honest-to-God ignition key which also fits the fork and seat locks. The ignition lock is incorporated into the left headlight mounting bolt.

There are two handlebar switches, each with a button. The left one is the horn (which should be louder) button and a three-position headlight beam switch (high, low, and flash). The right switch is the electric start button and the turn signal switch.

The headlight is turned on by the ignition switch, which has five positions: everything off; ignition off with rear parking light on; ignition on with no lights; ignition on with taillight; and ignition on with both lights.

Under the seat lies the 25-piece tool kit that includes tire irons, a tube patching kit, and a tire pump. The quality of the tools is very good, considering they are in a kit supplied with a motorcycle.

Another new feature is an unusual steering damper composed of a hydraulic cylinder between the bottom triple clamp and the frame and a large plastic knob atop the forks. The mount that attaches the damper to the triple clamp can slide from right to left, and turning the knob varies the mount's distance from the steering shaft. The knob has three positions: 0, 1, and 2. The "0" position places the mount closest to the steering shaft and provides the easiest steering; the "2" position places the mount furthest from the

BMW
BMW R 90/6
900cc

steering axis and provides the stiffest steering.

The telescopic front forks have internal springs and allow 6.25 inches of travel. Black rubber gaiters keep road crud away from the seals.

The rear shocks are manufactured by Boge and allow four inches of wheel travel. A built-in handle on the spring base allows three preload adjustments without any special spanner.

The engine is surrounded by a double-loop cradle frame. The seat, rear suspension, and rear accessories are supported by a bolt-on rear subframe. The housing for the driveshaft forms the right fork of the swingarm, and the driveshaft itself is connected to the transmission with a universal joint.

Center-pull spokes are laced to alloy wheels front and rear. A 3.25 x 19 Metzeler rib is mounted at the front, and a 4.00 x 18 universal Metzeler at the rear.

ENGINE AND GEARBOX: If the engine is cold you need the chokes to start the bike. The engine will fire off and idle immediately, but needs a few minutes warm-up before it will pull away strongly. The electric starter button requires a fairly hard push before it makes contact, but the starter won't operate if the clutch is engaged with the bike in gear.

The bike has a broad powerband with no sudden power surges through the rpm range. It accelerates smoothly from 1500 rpm in fifth gear, even with passenger aboard. At 60 mph the engine is turning about 3500 rpm, only halfway to the 7000-rpm redline.

Performance isn't generally considered to be the BMW forte, and compared to some other bikes of similar displacement, the R90/6 isn't a rocketship. BMW claims only a 14.0-second quarter-mile time for the R90/6, but our test bike logged 13.49 seconds at 96.8 mph, not bad for a tourer. We had a 1000cc Harley-Davidson Sportster along at the drag strip (for a test next month) and raced it against the BMW The BMW won, hands down.

Unless you get your weight well forward on the BMW, wheelies may be a problem if you are trying to get a quick hole shot—a fact that firmly impressed us during our first run at the dragstrip. Our tester wasn't far enough forward, and the tire spun for a few yards; when it got traction it almost looped the bike. The reason the BMW is prone to wheelie is that, in addition to normal rearward weight transfer during acceleration, the pinion on the driveshaft tries to rotate around the ring gear in the rear end and raises the whole bike when power is applied.

At 50 mph on level ground, you can accelerate and pass in fifth gear. If you are riding double or are on an upgrade, you must downshift into fourth for passing at 50, and if you're in a hurry, a quick jab down to third will give even more punch.

There is a large flywheel on the crankshaft to smooth out engine vibrations, and its inertia also helps low-speed torque. However, the additional inertia makes the engine less responsive to changes in throttle and load; because of the flywheel weight, the engine doesn't rev as quickly as, say, a four-cylinder bike. Nor does it return to idle immediately when the clutch is disengaged and the throttle closed. Because of this gradual response, it takes some time to learn how to shift without lurching. The faster the shift, the bigger the lurch when the clutch is released. The trick is to let the engine speed synchronize with the transmission speed before engaging the clutch after shifting.

In past years, one complaint about BMWs has been that the gearboxes clunk during shifting. Although the gearbox has been revamped this year, ours usually still clunked when it was shifted. By downshifting without the clutch—just blipping the throttle as we shifted—we could usually keep it from clunking. Normally, though,

we just used the clutch and let it clunk. The only shift that was usually silent was the upshift from fourth to fifth.

With a powerband as broad as the BMW's, a five-speed transmission hardly seems necessary, but it does bring the ratios closer together. They are evenly spaced and offer flexibility with however much power you want for any particular move. For example, you can come out of a 70-mph corner in any of three gears, depending on how much power you want. You can change the overall gearing on a BMW, but it's not cheap or terribly easy. You have to replace the ring and pinion set, which is a complicated job. Unless the bike is going to be used for something unusual, there is no reason to change the gearing.

The clutch lever required a fairly hefty pull, but otherwise the clutch was flawless. It never slipped or dragged, and the engagement was smooth and predictable.

HANDLING: The R90/6 is most at home with straight line cruising and casual cornering. It won't wobble at all on the freeway rain grooves, and if you are cruising in a straight line, you can take both hands off the bars and the bike will remain straight and steady. The only time it wiggles while going straight is if you make a hard shift with the power on; then it gives a slight twitch due to the offsetting torque of the engine and drive shaft.

Because the R90/6 is designed with long-distance touring comfort as the foremost consideration, some concessions are made in the area of high-speed handling. The soft suspension allows the bike to wallow slightly during fast cornering. We experimented extensively in one turn where it wallowed at 55–60 mph when pushed very hard. We were unable to make it wallow less with any combination of rear shock preload or steering damper position. The wallowing never turned into a tank-banging speed wobble or threatened to wrest control from the rider, but it made it difficult to take a precise line through those corners. But as long as you're not stuffing it through a turn at the absolute limit, the BMW corners precisely and predictably. You can pick a line with confidence in any corner, bumpy or smooth. It will also change lines willingly in the middle of a turn.

For comfort's sake, the suspension, especially the front forks, offers more wheel travel than most street bikes. However, overcoming centrifugal force while cornering causes much of the suspension travel to be used up, which reduces ground clearance. When cornering hard the side stand drags on the left and the brake pedal grounds on the right. If you lean still farther to the right, you can nick the valve cover, but that's leaning *very* far.

Part of the reason that the ground clearance is reduced during cornering is the torque reaction of the bike when the power is rolled off. The driveshaft tries to rotate around the ring gear and push the motorcycle down during deceleration; and applying throttle will cause the driveshaft to raise the motorcycle. The result is that acceleration means more ground clearance, and backing off the throttle means a reduction in ground clearance. Most riders will never be concerned by the resulting dragging of the sidestand and brake pedal, but riders who corner fast will have to deal with it.

Because of the low center of gravity offered by the horizontal engine, it's a breeze to flick the bike from side to side in an S bend. The lower center of gravity also makes the bike seem lighter when you are supporting it at a stop light. And you don't have to be built like King Kong to back it into a parking place.

The Metzeler tires are designed for long life at high speed. The rubber compound is fairly hard, but the tires hold well within the limits imposed by the bike's power and ground clearance.

COMFORT AND RIDE: One of the things that BMWs are most famous for is comfort. They may give away something in handling, but they give away almost nothing in comfort.

The seat is long, wide and comfortable. Long hours in the saddle won't make your posterior sore. The bars are rather narrow (27.7 inches) but are wide enough for easy maneuvering at low speed. The narrowness is good for high-speed comfort and touring because it keeps you from being spread out in the wind. The Magura grips are large and soft enough that your hands won't get cramped.

A friction damper on the twistgrip allows you to ride long distances without constantly having to twist the throttle. Turning a thumb screw pushes a friction shoe against the drum of the twistgrip, but you can still close the throttle by forcibly turning it back.

The only part of you that will ever get sore while riding the R90/6 will be your left hand, due to the heavy-pulling clutch. But that will only occur on roads where you must shift frequently.

Because the right cylinder is offset to the rear in relation to the left cylinder, the right footpeg is equally offset behind the left footpeg. Nobody noticed this unless it was pointed out to them. Two of our testers were bothered by the right intake tube, which occasionally hit their ankles.

The seat-bar-footpeg arrangement fits most riders. You lean forward on the bars only a little bit, which works out fine once you're moving and there is wind pressure on your chest.

The front suspension, with its long travel, absorbs big jolting bumps and tiny road cracks equally well. The rear suspension never pounds a solo rider, even with the shocks on their stiffest setting; thus your back doesn't get sore after a day of riding. For the average rider, the best ride is with the shocks on the softest position; the best position for hard cornering is the stiffest position.

The R90/6 vibrates far less than most bikes. Our test bike was smooth at highway speeds, but during hard acceleration and when the engine was under load, it had an annoying low-frequency vibration.

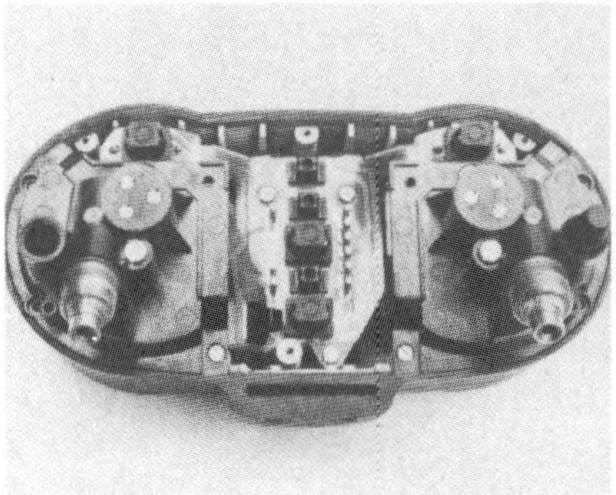

BMW has eliminated bird nests in the instrument wiring by using printed circuitry.

The BMW front disc is very thin, and there are various advantages to mounting the caliper behind the fork leg. Notice the center-pull spokes.

Because the cylinders are offset, the footpegs are offset, but it's not easily noticeable. The Bing carburetors get air from an airbox at the top of the engine cases.

The master cylinder for the front brake system is under the tank and is protected from damage in a crash. A warning light tells when fluid is low.

And it was noticeable at any rpm, although not enough to cause fatigue.

Noise can also be fatiguing, but the very low level of exhaust and mechanical noise from the R90/6 makes a ride to work or a tour through the country a really pleasant experience for both you and those around you. BMW has set an enviable standard of excellence in this area.

The passenger is well taken care of; he has plenty of room on his part of the seat and the passenger pegs are adjustable. Instead of a grab strap, BMW uses a hand rail around the back of the seat for those not inclined to hold onto the rider.

THE BRAKES: Feedback and feel from the front brake were excellent, and the brake required a great deal of force to lock the wheel. There was no noticeable pull to either side.

The rod-operated rear brake is too powerful and insensitive. It can be locked easily, especially if the rider downshifts while braking. Downshifting into first gear while braking would often lock the rear wheel. The best stopping distance at 60 mph was 137 ft. 6 in.

Although the brakes were quiet when cold, they squeaked when they got hot. The front brake emitted a soft, high-pitched squeal, and the rear shoes squeaked louder but with a lower pitch.

Instead of an external pointer to meet Federal demands for brake wear indicators, the BMW has two viewing slots in the rear hub. You look through these slots to check the condition of the brake shoes. We thought the slots might allow water to enter the drum and make the brake ineffective, so we thoroughly doused the rear brake with a hose and immediately went for a ride. The rear brake was as strong as ever.

RELIABILITY AND MAINTENANCE: One of the most important factors in deciding a bike's worth, and the hardest to measure in a road test, is reliability. A bike can be a rock-steady rocketship that stops in an instant, but if it self-destructs after 5000 miles it had better have a pretty low price tag to make it worth considering. Owners of any particular motorcycle always tend to be a little optimistic about its qualities because their egos are involved, but we called one mechanic friend who has been working on BMWs for a few years and asked him how often he had to do top-end jobs and replace crankshafts.

There was a pause. "I've never done a top-end job on one," he said. "And I don't even know what a BMW crankshaft looks like."

Was there any problem that they commonly had? "Yeah," he said, "they turn pipes blue."

That *was* only one mechanic, but we've never heard anyone say anything different; the legendary BMW reliability comes from more than the overconfidence of a few happy owners. The new engine size introduces a variable to that legendary

SPECIFICATIONS

Engine type	four-stroke
Cylinder arrangement	horizontally-opposed twin
Valve arrangement	overhead valve, pushrod-operated
Bore and stroke	90mm x 70.6mm
Displacement	898.2cc
Compression ratio	9.0:1
Ignition	battery/dual coil/single point
Charging system	12-volt AC generator, voltage regulator, diode
Carburetion	two 32mm Bing constant-velocity
Air filter	disposable paper element
Lubrication	wet-sump, 2.4 qt. sump capacity
Primary drive	gear, 1.5:1
Clutch	dry, 1 drive plate, 1 driven plate
Starting system	electric and kick, in neutral only
Transmission	5-speed, left foot shift
Overall drive ratios	(1) 13.6; (2) 8.83; (3) 6.4; (4) 5.16; (5) 4.63
Pinion gear	11-tooth
Ring gear	34-tooth
Front forks	telescopic, 6.25 in. travel
Rear shocks	Boge, 3-way adjustable, 4 in. travel
Front brake	single-action hydraulic caliper, 10.2 in. disc
Rear brake	drum, single-leading shoe, rod-operated
Front tire	3.25 x 19 Metzeler rib
Rear tire	4.00 x 18 Metzeler touring
Frame	tubular steel, double cradle, bolt-on rear subframe
Wheelbase	59 in.
Length	86.6 in.
Weight	468 lbs.
Weight distribution	45.5% front, 54.5 % rear
Ground clearance	6.1 in., at sidestand
Seat height	32.5 in., unladen
Handlebar width	27.7 in.
Handlebar grip height	41.4 in.
Footpeg height	11.5 in. right, 11.1 in. left
Instrumentation	tachometer, speedometer, odometer, tripmeter resettable to zero.
Gas tank	steel, 4.8 gal.
Gas mileage	46.8 mpg average
Best ¼-mile acceleration	13.49 sec., 96.8 mph
Stopping distance from 30 mph	35 ft. 1 in.
Stopping distance from 60 mph	137 ft. 6 in.
Suggested price	$2930 East Coast, $2950 West Coast.

BMW R90/6

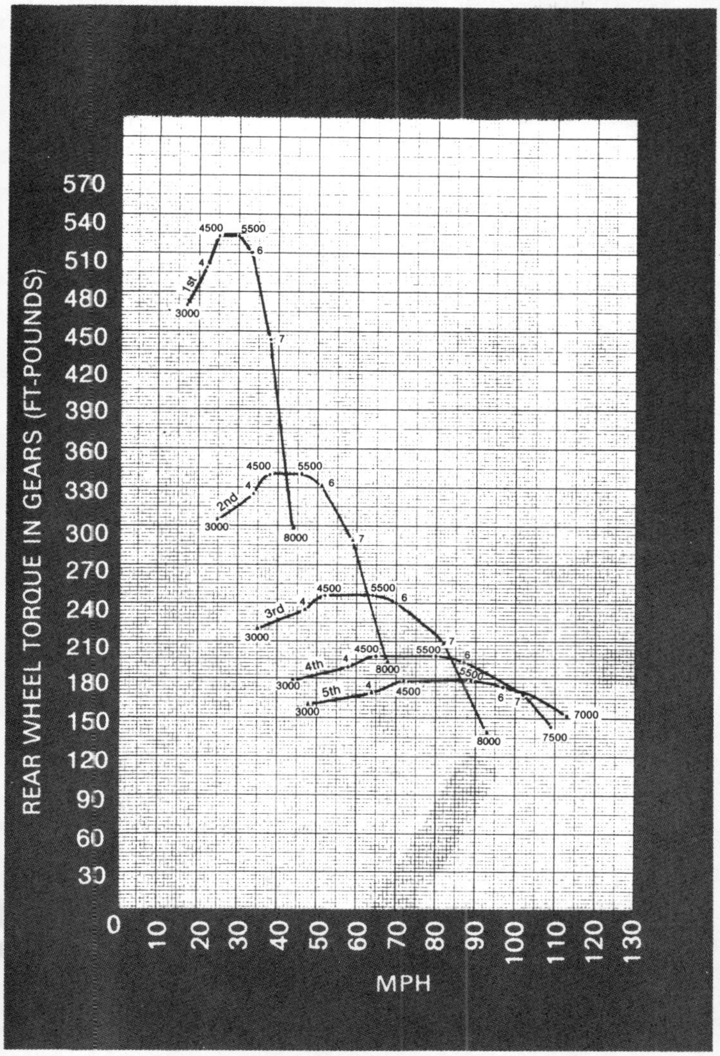
REAR WHEEL TORQUE IN GEARS (FT-POUNDS)
MPH
1st
2nd
3rd
4th
5th
0 10 20 30 40 50 60 70 80 90 100 110 120 130
30 60 90 120 150 180 210 240 270 300 330 360 390 420 450 480 510 540 570

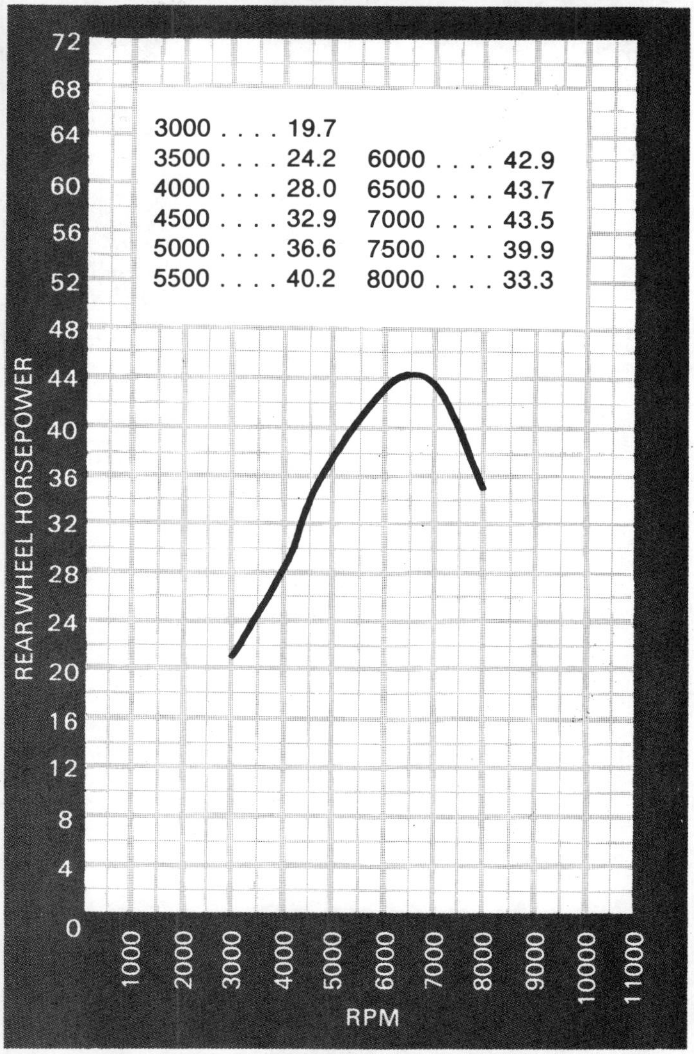
REAR WHEEL HORSEPOWER
RPM
3000 . . . . 19.7
3500 . . . . 24.2
4000 . . . . 28.0
4500 . . . . 32.9
5000 . . . . 36.6
5500 . . . . 40.2
6000 . . . . 42.9
6500 . . . . 43.7
7000 . . . . 43.5
7500 . . . . 39.9
8000 . . . . 33.3
0 4 8 12 16 20 24 28 32 36 40 44 48 52 56 60 64 68 72
1000 2000 3000 4000 5000 6000 7000 8000 9000 10000 11000

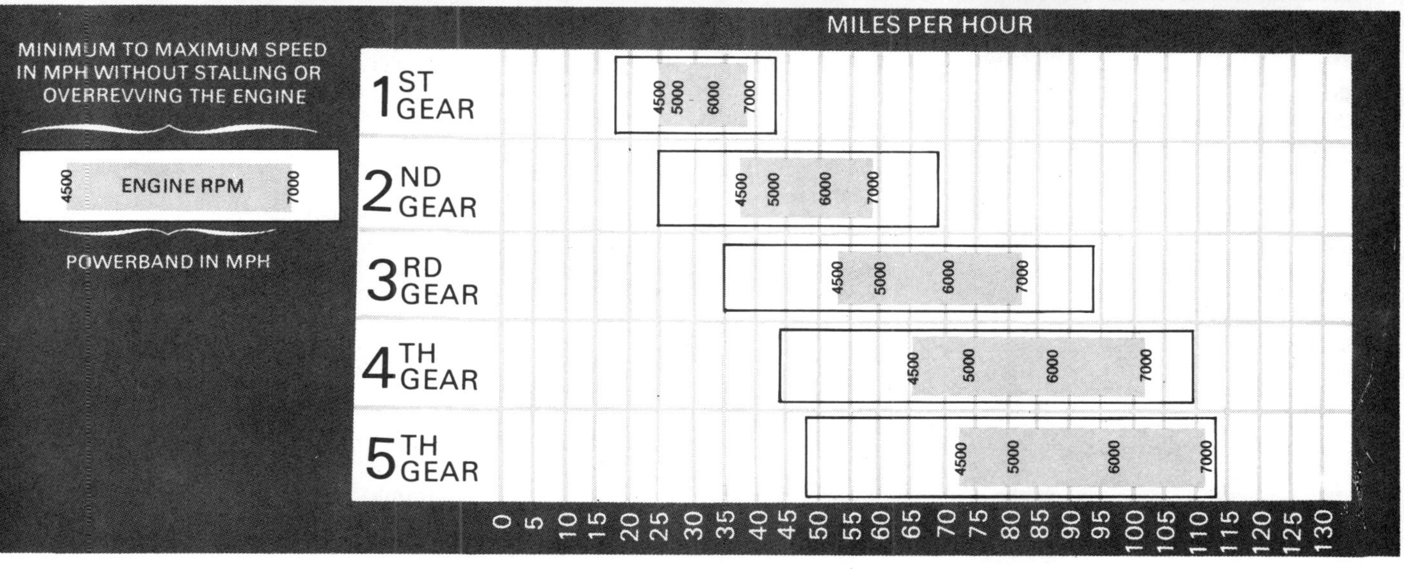
MINIMUM TO MAXIMUM SPEED IN MPH WITHOUT STALLING OR OVERREVVING THE ENGINE
4500 ENGINE RPM 7000
POWERBAND IN MPH
MILES PER HOUR
1ST GEAR
2ND GEAR
3RD GEAR
4TH GEAR
5TH GEAR
4500 5000 6000 7000
0 5 10 15 20 25 30 35 40 45 50 55 60 65 70 75 80 85 90 95 100 105 110 115 120 125 130

reliability, but we don't feel that the bore-job will affect engine life. BMW engineers have proven themselves more careful than that.

Commonly maintained parts are easy to reach: The valves, carburetors, points, air cleaner and large 12-volt battery are all at hand.

Two maintenance benefits of the drive-shaft are: (1) It only requires an occasional change of lubricant; (2) you never have to align the rear wheel.

The spokes are of the center-pull variety and require less attention than conventional angled ones which are found on most bikes.

During the test we never had to do anything more than add gas and oil. We didn't even have to adjust the clutch or brakes after acceleration and brake testing.

MISCELLANEOUS: The Bosch quartz iodine headlamp is excellent. It throws a bright, wide beam that thoroughly lights a large area.

Two mirrors are supplied with the bike. The right one on our bike badly distorted the images over much of its area.

The speedometer includes a resettable tripmeter, something that is useful for telling how long it's been since you last filled the gas tank. It was about four mph optimistic at 60 mph.

The BMW sidestand snaps back automatically when the bike isn't leaning on it. This might be overlooked by someone unfamiliar with the bike; he may lean the bike upright and then try to lean it back over onto nothing. A centerstand is also standard equipment.

There is a helmet hook located beneath the seat. You can slip the D rings of your helmet on the hook and then lock the seat and leave your helmet with the bike. Unfortunately, a healthy tug will open the seat whether it is locked or not. And while someone is stealing your helmet, he'll probably take your tool kit and tire pump too. The lock needs redesigning, along with a repositioning of the helmet hook. If you try to carry a helmet on the hook while you're moving, the helmet hits the rear tire.

We averaged 46.8 miles per gallon during the test, or almost 225 miles per tank. We got one bad batch of "premium" gas that made the engine ping when the throttle was open wide at low rpm. If the quality of gas seems to be low in your area, or you plan on touring in areas where high octane fuel isn't available, your local BMW dealer can show you how to retard your timing to suit the fuel.

The only problems we had during the test were improperly adjusted carburetors (which we adjusted and had no further problem with) and very slight oil seepage around the case joints when the speedometer registered 5200 miles. And the pipes were blue, as our mechanic friend mentioned.

SUMMARY AND CONCLUSION: The BMW R90/6 is a powerful motorcycle, designed to compete in the performance-conscious market of 1974. Basically, the R90/6 is a bored-out, snappy 600cc tourer, but by restyling, adding a front disc brake and several other innovations and features, BMW has created a strong competitor in its class.

The R90/6 accents touring, but still maintains an identity as a sporting machine. The engine is powerful enough for good high speed performance and yet remains absolutely tractable at low speed. The five-speed transmission clunks but has well-spaced ratios; the clutch is smooth, but heavy-pulling.

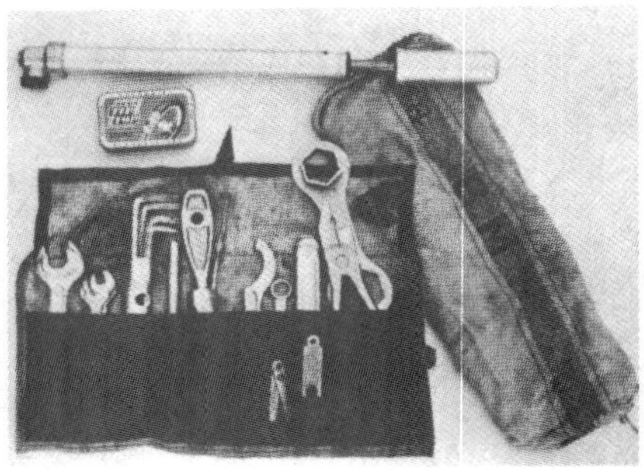

The tool kit is excellent. The 25-piece kit includes tire irons, a pump, a tube patch kit, and a monogrammed polishing cloth.

The R90/6 is very much at home on the highways and interstates. It is smooth, stable, and quiet at touring speeds but some of the concessions to comfort and ride, so valuable in a touring situation, limit it somewhat during hard cornering. Routine maintenance is kept to a minimum, and what maintenance there is can be done quickly.

As a tourer, the R90/6 is not outdone by any motorcycle, but some of the Japanese multis have closed the gap on BMW in many areas where BMW used to stand alone. And although the prices of Japanese bikes are rising, they are still cheaper than the $2950 price of the R90/6.

So, in the end, the R90/6 will have to sell itself by reputation, reliability, charisma, and unique features that have appeal to some buyers and are unavailable elsewhere. ***CG***

Cycle World Road Test

BMW R90S

Overlooking The Superlatives Found On The R90 Sport Is Difficult... And Overlooking The Price Is Impossible.

■ IT IS surprising, to be sure, and yet, BMW's new R90S, despite a price tag of nearly $3500 (which means an out the door price of over $3800 in most areas), is being scooped up in remarkable numbers by a lot of buyers. Butler and Smith, BMW's U.S. Importer, expects to sell each and every Sport model it can bring into the country this year, and that's saying a hell of a lot.

It tells you that the BMW image of yesterday (stodgy, conservative, unexciting), has at last departed the scene. It tells you that this particular motorcycle must offer a tremendous amount of appeal and satisfaction to make working class people dig deeper into their pockets than they ever have before just for the privilege of riding home on two wheels. It tells you that the marketing staff at BMW knew what they were doing when they typed out their list price sheets. And it lets you know in no uncertain terms that people are more than willing to pay dearly for what they want. Obviously there are many motorcyclists out there who very dearly want a BMW R90 Sport. And we don't blame them for a minute.

Just what is it that makes the Sport so special, exceptional and remarkable? A combination of everything, actually. Because here is a motorcycle that offers plenty of power, which delivers not only superbike speed and acceleration, but operational economy as well. Simplicity? Yes, that too and reliability besides.

Did we hear you mention comfort? You'll be hard pressed to find a machine that delivers more, including smoothness and quietness, than the Sport offers. On top of everything it handles fairly decently, stops on the proverbial dime, stays cleaner than a chain driven bike and has a host of conveniences. Yup. Resale value too. And on top of everything, styling is superb; right on the verge of being a traffic stopper. But don't get the idea that it's perfect. It isn't.

Mechanically, the Sport is very similar to the more conventionally styled R90/6, which we tested earlier this year (Cycle World, Feb. '74), but is styled to coincide with its sport/touring concept.

Both the R75/6 and R90/6 carry a single disc brake at the front, but the 90S goes them one better with two units. Caliper mounts are cast into the rear sides of the alloy fork legs where the metallic pucks grasp the greater-than-10-inches-in-diameter stainless steel discs upon brake application.

Activation is via a hydraulic master cylinder that is not located on the handlebars as is usually true of bikes using this method of braking. Instead, BMW has located the unit under the 6.4-gal. fuel tank, where it not only stays out of sight and cleans up styling a bit, but is less likely to be damaged if the machine were to be knocked over in a parking lot. A master cylinder doesn't need frequent checking, so its out-of-the-way position isn't detrimental in that respect.

Further up the fork legs, the absence of rubber boots is apparent if you're at all use to other BMW models. This way, front end appearance is not so busy and the machine is given a "lighter" look. Fenders are fiberglass, as is the snappy-looking head fairing and seat tail section.

These pieces, along with the steel fuel tank, are finished in a custom fogged-gray color, which blends from a pearl silver into a gray black. A clear lacquer is applied over this to add depth, then the gold striping goes on. Only thing is, this pin striping has been applied with gold *tape*, not gold *paint*. When we saw this we had a fit; a $3500 motorcycle and the striping is out of

Photography: Bob Atkinson, D. Randy Riggs

a J.C. Whitney catalog! Must be a reason for it, so we checked.

Seems as though BMW couldn't find a suitable paint for the job. The ladies at the factory who do the hand striping weren't able to make the paint look uniform enough, so tape it was. Later models may get the paint if they find the right stuff, but until then, it'll have to be tape.

The head fairing that adds such a racy appearance to the machine, functions surprisingly well, in spite of its size. Above 70 mph it really takes much of the wind blast off the upper portion of the rider, and that can mean a lot in terms of comfort after a long day's ride. Another bonus is the inclusion of an instrument nacelle that holds both an ammeter guage and an electric clock, an item that hasn't been installed on a motorcycle as standard equipment in more than 25 years.

Below this panel, the new instruments found on /6 series BMWs rest in the normal location. Non-glare faces on both the tach and speedometer are a help; and somehow BMW has come up with warning lights that are actually visible in sunlight, yet aren't too obtrusive or distracting at night. The ignition switch resides in the left side of the headlight and has positions for parking lights, ignition activation and normal lighting.

The headlight throws an extremely bright light beam down the road at night, but the flat beam pattern doesn't work too well in cornering. Handlebar switches are easily operated with the thumb of either hand; pushing in the left switch blasts the horn, depressing it downward flashes the high beam and a click stop upward turns the high beam on continually. On the right the switch operates in the same manner, but it operates the starter and turn indicators.

Magura control levers are a nice touch, but the throttle assembly (another Magura item), requires too much of a twist for wide open running. To really wick it up, the rider has to grab two handfuls instead of one. But one bonus of the mechanism is a throttle friction screw, so the rider can set the engine at a cruising speed and take his hand from the grip. It really helps on a long haul.

Frame design is typically BMW and the Sport possesses the same unit as the R75 series, but with additional gussets and a longer swinging arm added. New this year on each of the BeeEm models is a rather unique steering damper assembly. The unit is hydraulic and double-acting and has three positions, set with a twist of the wrist by the rider. At one end the shock absorbing unit connects to the frame. Opposite this, the end fits in a slotted rectangular box that locates under the lower triple clamp. By changing positions of the knob, the rider moves the unit through various degrees of the steering axis. Damping in turn goes from neutral to stiff; it's a surprisingly nice set-up.

Although the Sport's suspension softness contributes greatly to a comfortable ride over rough roads, it can be annoying when trying to snake your way at high speed on a winding road. Front fork travel is overly abundant at eight inches; this length coupled with soft springing allows the R90 to wallow at speed.

Drive into a corner hard and get on the brakes and the Sport will nosedive alarmingly. This dive uses up important ground clearance when you're ready to pitch it into the turn. But if the road gets rough, the advantage is yours. It depends on the situation. Were the Sport our own, however, we'd perform some fork modifications.

At the rear, we offer the same criticism that we did on the front. For really fast riding, the damper units are simply too soft, even in their stiffest positions. But, in turn, they allow one of the most comfortable rides imaginable.

If the seat looks as though it does the job, you're right. Both rider and passenger are amply cradled. Too, there's a grab bar for the hanger-on, and the unit flips up. Underneath there are two large trays for tools or a camera...whatever. The front tray lifts out to reveal the 12-volt, 25-amp-hour battery.

As usual, the tools supplied are more than enough to do plenty of maintenance or almost any roadside repair; there's even a tire patch kit, an air pump and a rag to wipe your

PARTS PRICING

Warranty	6 mo./6000 miles
Major Tuneup	$50.00 plus parts
Air Filter Element	$7.50
Rear Tire (standard)	$33.38
Headlight Bulb	$10.95
Taillight Bulb	$.70
Turn Indicator Bulbs	$.50
Clutch Cables	$5.55
Throttle Cables ea.	$6.00
Brake Cables	$5.45
Ignition Parts	
Points	$3.95
Condenser	$3.10
High Tension Coil, ea.	$19.85

hands. BMW thinks of nearly everything, including a lock to keep thieves out and a prong to hold a couple of helmets when they're not in use.

Like other BMWs, the Sport features both a side and center stand; the side stand is self-retracting, something we're not that keen on, but the center stand is about the easiest to use we've seen on any motorcycle.

The typical BMW niceties are naturally found on this top-of-the-line model. On top of the others we already mentioned that includes aluminum wheel rims. Tires are by Metzeler, and though their hard compound offers exceptional tire life, traction in fast cornering is borderline acceptable. If you're a tourer at heart, leave the Metzelers alone. But if you get your kicks by boogeying down a mountain pass, better fit some TT100s or something similar.

Another strong point of owning and maintaining a BMW is the overall simplicity of design. Though the engine displaces nearly 900cc, there is no mechanical complexity. Carburetion is not found in rows and neither are cylinders. Instead, the flat Twin juts a cylinder out each side of its unitized central housing. This not only sticks those finned aluminum bores out into a cooling airstream, but it allows a mechanic an unhindered access to valve adjustment and each 38mm Dell'Orto carburetor. And that makes a bunch of difference when you're doing the work yourself at home.

Both the crankshaft and I-beam rods are forged, and the crank in the Sport is the same found in all three BeeEm engines. It rides in three plain-type automotive main bearings, with rod journals spaced 180 degrees apart. Pistons are large enough to look as though they've come from the internals of a mining locomotive. Compression ratio is up to 9.5:1 on the Sport, from the /6's 9.0:1. This is the only internal difference between the two 900s; the rest, with the exception of the rear axle ratios, is outside.

Early production R90s had no choke system on the giant 38mm Dell'Orto carbs; just the messy float ticklers. Fortunately, BMW didn't go that route for long, and a conventional choke system for cold weather starting is now employed. Intake air is drawn through a paper filter element hidden in the hollow aluminum crankcase hump. Not only does this system eliminate some of the dirt before it even reaches the filter, but mechanical and intake noises are held down, as well.

BMWs have long been due for a decent five-speed gearbox; this year it finally happened. But the unit on our Sport test machine didn't work nearly as well as that on the 90/6 model we tested earlier in the year. Dissappointing, considering the overall performance of the rest of the machine.

Neutral was difficult to find the majority of the time when stopped at a light; trying to find it when rolling to a stop didn't help much either. Gear changes were not the smoothest, and clutch pull was not the lightest. Upsetting, indeed, on a $3500 motorcycle. The S transmission is the same unit used on other BMWs, ratios included. But the Sport's high rear axle ratio (3.00:1 versus 3.09:1 for the /6 900), requires some clutch slipping in town, emphasized by the fact that the S engine is not that thrilled with the prospect of accepting much throttle below 3000 rpm. As a result, in-town riding is more pleasant on the roadster version of the 900.

And the hardest decision that a prospective BMW 900 buyer would have to make would be which 900 to ride home on. The roadster version does nearly everything that the S version will do; it's just a tad slower and lacks the styling punch and image that the Sport carries. But the Sport is another $500, and regardless of how fine a motorcycle it is, that extra tab tacked on to an already very expensive piece of machinery is more than many people can handle.

BMW R90S

SPECIFICATIONS

List price	$3430
Suspension, front	telescopic fork
Suspension, rear	swinging arm
Tire, front	3.25-19
Tire, rear	4.00-18
Brake, front, eff. dia. x width, in.	(2) 9.8 x 1.34
Brake, rear, diameter x width, in.	7.87 x 1.18
Total brake swept area, sq. in.	182.5
Brake loading, lb./sq. in. (160-lb. rider)	2.66
Engine, type	four-stroke, horizontally-opposed Twin
Bore x stroke, in., mm	3.54 x 2.78, 90 x 70.6
Piston displacement, cu. in., cc	54.78, 898
Compression ratio	9.5:1
Claimed bhp @ rpm	75 @ 7200
Claimed torque @ rpm, lb.-ft.	55 @ 5500
Carburetion	(2) 38mm Dell'Orto
Ignition	battery
Oil system	pressure, wet sump
Oil capacity, pt.	4.7
Fuel capacity, U.S. gal.	6.3
Recommended fuel	premium
Starting system	electric, Kick
Lighting system	12-V alternator
Air filtration	dry, treated paper
Clutch	single, dry plate
Final drive	Palloid pattern crown wheel and pinion
Gear ratios, overall: 1	
5th	4.50
4th	5.01
3rd	6.21
2nd	8.58
1st	13.20
Wheelbase, in.	57.7
Seat height, in.	32.0
Seat width, in.	11.0
Handlebar width, in.	24.0
Footpeg height, in.	11.75
Ground clearance, in.	6.5
Front fork rake angle, degrees	N.A.
Trail, in.	3.5
Curb weight (w/half-tank fuel), lb.	486
Weight bias, front/rear, percent	46/54
Test weight (fuel and rider), lb.	664.9
Mileage at completion of test	3433

TEST CONDITIONS

Air temperature, degrees F	81
Humidity, percent	54
Barometric pressure, in. hg.	29.90
Altitude above mean sea level, ft.	383
Wind velocity, mph	9-12
Strip alignment, relative wind:	

PERFORMANCE

Top speed (actual @ 7000 rpm), mph	123
Computed top speed in gears (@ 7300 rpm), mph	
5th	123
4th	110
3rd	89
2nd	64
1st	42
Mph/1000 rpm, top gear	16.8
Engine revolutions/mile, top gear	3578.40
Piston speed (@ 7300 rpm), ft./min.	3382
Lb./hp (160-lb. rider)	8.61
Fuel consumption, mpg	45
Speedometer error:	
50 mph indicated, actually	45
60 mph indicated, actually	56
70 mph indicated, actually	64
Braking distance:	
from 30 mph, ft.	30
from 60 mph, ft.	125
Acceleration, zero to:	
30 mph, sec.	2.8
40 mph, sec.	3.8
50 mph, sec.	4.9
60 mph, sec.	5.7
70 mph, sec.	7.0
80 mph, sec.	7.8
90 mph, sec.	9.0
100 mph, sec.	11.9
Standing one-eight mile, sec.	8.8
terminal speed, mph	83.02
Standing one-quarter mile, sec.	13.05
terminal speed, mph	103

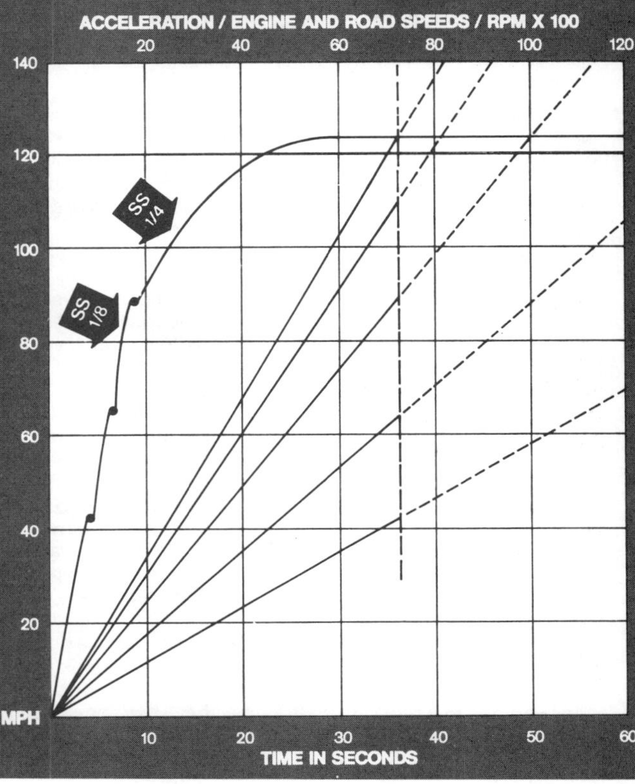

BIGGER, BETTER AND BMW

R90/6 —easier handling and tall performance get BMW's big motor close to justifying the cost

BMW's official reason for not changing a concept which goes back at least to the early 'twenties, is that from an engineering point of view they cannot improve on it.

They may be right. Having ridden all three models in the stroke-6 series, I have to admit that I can't find anything wrong with them. Of course, you can find other machines which will do the same things from performance, handling or comfort standpoints. But not many manage to combine all these attributes quite so well — and it's when you need to do a really hard run that you appreciate this.

The R90 covered 2300 miles in six and a bit days riding time. I'm not trying to claim any records, in fact although the BMW was ridden hard and fast, it was all conducted in a most leisurely manner.

As it is this kind of riding which seems to be BMW's strongest point, we thought we'd do a bit of grand touring. I took the R90 across France and back, carrying a payload of about 350lb including wife and luggage.

We deliberately stayed away from the main routes which follow valleys and low-lying land and chose the hilly regions instead.

And yet five or six hours riding would still see three hundred miles behind us. That's the way it went — we'd ride for four, five or six hours with a couple of coffee breaks and stay the night. The BMW was always ready to go and never really seemed to be exerting itself. Every 200 miles it would run on to reserve, and that's about all we heard from it.

I took a notebook with me, intending to enter up all the events as they happened and come back with a blow-by-blow account. Looking at the notes now, apart from an early entry about panniers and tyre pressures, all you see is mileage readings, quantities of fuel and more mileage readings.

That, I suppose is what BMW-ing is all about — fast, smooth, reliable and with a maximum of comfort or if you like, a minimum of effort.

And comfort is the important factor. I can't think of any other machine which gives such an easy ride and still handles as well as the BMW, even with a full load on board.

Room for the load caused the first problem. We decided to use Craven panniers and top-box with their carrier which is designed for the BMW's. Now, like most other manufacturers, BMW are wary of such things. They say, in the owners handbook that panniers should not weigh more than 15lb each when loaded and anything on the luggage rack (or top box in our case) should be kept as light as possible. They go on to recommend that luggage be carried on the fuel tank and I believe they supply a pannier type bag which goes on the tank.

However we did not have such equipment, we also wanted to test the Craven panniers and this seemed to be an ideal time to do it.

The reason manufacturers don't like loads being carried on the back of the machine is that they can upset the handling drastically. We had the panniers fitted, I loaded them up in a haphazard sort of way, just stuffing things in where they would fit, and set off down the road. As the speedo went through the 35 mph mark the BM's handlebar went into a really violent wobble. Above or below the magic 37 or 38 mph the wobble disappeared but just at that speed even the hardest steering damper setting

BMW R90/6

couldn't contain it.

I'd never had a wobble which could be so easily and predictably produced and the subsequent experiments lead to some very interesting conclusions — especially as the subject of speed wobbles has caused great controversy ever since the first guy fell of a motorcycle.

Anyway the story behind all that will have to wait until next month, the immediate problem was to eliminate the wobble and get the BMW's stability back to normal.

A check over the machine revealed no faults, the carriers were properly fitted; increasing the tyre pressures to 35 psi to match the extra load and moving the rear suspension up to hardest subdued the wobble slightly but not enough. Then I came across BMW's recommended loads in the handbook. Out came the bathroom scales, and I found that I'd not only got twice as much in the left hand pannier as the right one, but all the heavy items seemed to have found their way into the top box.

After some careful redistribution I got the panniers weighing 16lb each (only 1 lb over the limit) and the top box load down to 9 or 10lb. That just about did it, the wobble was still there, but only as a tremor, it would only build up into a real wobble if the rider aggravated it by holding wobble speed and letting go of the handlebar.

Once I'd appreciated how critical the load distribution was, the panniers proved an excellent accessory. There is a choice of a permanent fitting or QD panniers which are secured by a Dzus fastener. I chose the latter as these have a handle and can be used virtually as suitcases. Although they don't appear to be very big they certainly hold a lot of stuff and they stayed completely weatherproof.

The top box is also very spacious, and it's tempting to overload it because of this. It turned out that anything approaching a heavy load in the top box was the major cause of the wobble — probably because this was mounted furthest from the wheelbase and had a greater leverage or moment of inertia.

Both panniers and top box are lockable and the only snags were again with the top box whose lid didn't always want to fit and which mysteriously leaked when the machine was ridden through a rainstorm.

Back on the R90 I'd run up nearly 200 miles in the UK before arriving at Ramsgate's Hoverport. I'd decided that the Hoverlloyd service across to Calais would be faster than going by boat, and as motorcycles can usually be squeezed on to hovercraft (or ships for that matter) with no problems, we had an open ticket with no reservation for a particular flight. What I didn't know was that they load the devices up maybe half an hour before they leave and that motorcycles have to go on first. So we had to wait for the next one . . . something that's worth bearing in mind if it's important to catch a particular flight.

Out of Calais the first thing that I noticed was how much faster the French roads seemed to be. Just over four gallons and some 220 miles later the BMW ran on to reserve — and it was only a little over three hours since we'd left Calais!

We were averaging nearly 70mph, with as little effort as it would take to average a mere 50mph in this country. I put it down to the fact that everyone in France drives flat out all the time and I must have been subconsciously joining in. With a more objective point of view from the pillion seat, Liz pointed out that there are no hedges alongside French roads and that you can see all the road all of the time — there are very few blind corners, so you just don't need to slow down so much or so often.

On top of that continental motorists are much more aware of motorcyclists — they actually use their mirrors, and invariably pull over, waving the motorcycle past or giving a couple of blinks on the nearside indicator.

If the BM's speed and comfort allowed it to make use of these favourable, if unfamiliar, conditions, the French roads certainly put brakes and handling to an exacting test.

From a long, smooth straight, I'd be lining the /6 up for what I'd judged to a 40 mph corner, braking down from about 90 mph when right on the entry to the corner there'd be the dreaded triangle with a vertical black bar and the legend "chaussée déformée". Very quickly I'd have to scrub off another 20 mph before the BM was slithering and kicking over a mess of bumps, ruts, potholes and melting tar.

In fact all French road signs tend to come back to the same subject. You don't need to be a linguist, they all mean "here come something which makes people fall of motorcycles". There's one variation, the silhouette of a mountainside with lumps falling off it, which I take to mean "if you don't fall of the mountain, the mountain will fall on you" and now I know how Mahomet might have felt . . .

It says something for the R90's brakes and handling that it got through all this unscathed except for a small portion ground off the stand where it

The hydraulic steering damper is joined to a link which can be moved outwards to increase the stroke.

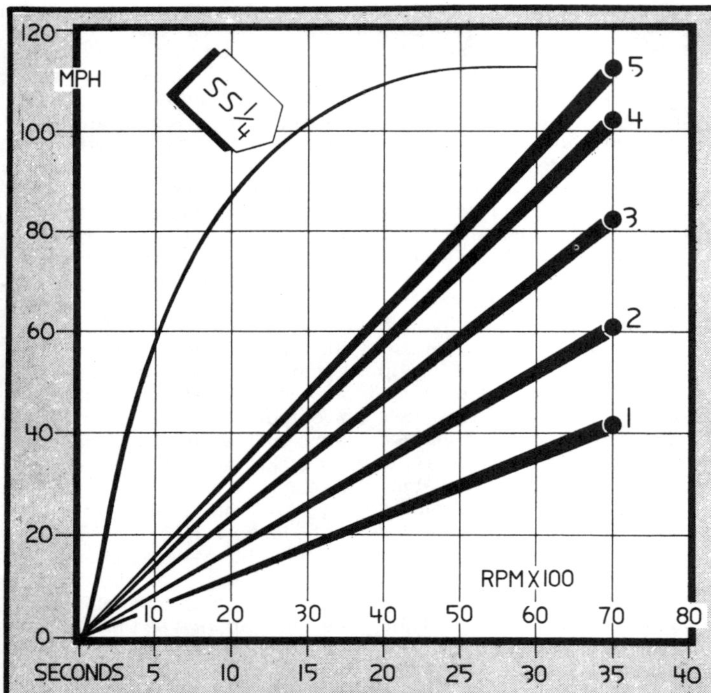

SPECIFICATION

braking distance from 30 mph 29ft.
fuel consumption, average 53 mpg

engine type	OHV horizontally opposed twin
displacement	898 ccm
bore x stroke	90 x 70.6 mm
compression ratio	9.0:1
claimed output	67 hp SAE at 6700 rpm, 60 hp DIN at 6500 52.8 lb ft torque at 5500 rpm
lubrication	wet sump, pressure feed from Eaton pump
carburettor	two 32 mm Bing CV
ignition system	twin cb and coil
lighting system charging system	12V ac/dc from 3-phase alternator 280W via diode rectifier and regulator to 25 ah battery
clutch	single plate, diaphragm spring
gear ratios,	4.40, 2.86, 2.07, 1.67, 1.50:1
primary drive	direct to clutch, reduction gear to gearbox input shaft
secondary drive	enclosed shaft with palloid crownwheel and pinion. Reduction 3.09 (optional 3.2 3.0 or 2.91)
mph/1000 rpm in top	16.3
fuel tank capacity	5.3 gal inc .44 gal reserve
oil tank capacity	4 pints
tyres, front	3.25 x 19 Metzeler
tyres, rear	4.00 x 18 Metzeler
brakes, front	10.2 inch disc, cable/hydraulic operation
brakes, rear	7.9 inch sls drum, rod operation
wheelbase	57.7 in
ground clearance	6.5 in
seat height	31.9 in
dry weight	436 lb
test weight	463 lb
suspension front	two-way damped telescopic fork
suspension rear	swinging arm, 3-position two-way dampers
trail	3.5 in
speedo error	approx. 5 per cent fast
list price	£1520.85 inc. VAT
manufacturer	BMW GB, 991, Great West Road, Brentford.

had grazed the floor.

And it was still turning in the high averages. On stretches where I deliberately drove it hard, or mountain roads where we were continually braking hard and using the gears, fuel consumption dropped to 50 mpg. On easier sections, cruising around 90mph, we were getting nearly 60 mpg.

Cruising speed was only limited by the mounting discomfort of the wind pressure. A big handful would take it easily up to an indicated 110 mph, and once or twice I saw 115, under favourable conditions.

A rough check indicated that the speedo was about 5 per cent fast, so bearing in mind that we were fully laden, sitting upright, this tallies with BMW's claim of 111mph.

I imagine the R90S might be a bettter machine for this kind of riding. It has a higher top gear and the flyscreen fairing would certainly improve comfort over the 90 mph mark. There is, incidentally, an alternative, high ratio crownwheel and pinion available for the R90.

Having sung the /6's praises — I still can't get over how *easy* it all was — I should add that it's a vast improvement over the earlier /5 models.

The biggest differences must be the handling — there's a longer swinging arm and slightly less suspension movement — and the gears. Having five gears gives closer ratios and the gearshift only clunks between first and second, or if you change gear at low engine speeds. You can also downshift while the /6 is cranked over — something which used to produce much leaping about on the /5s.

BMW's disc brakes all have a tendency to squeal — this was no exception, but as the pads wore it somehow cured itself. Apart from that they work really well, except in the wet when they take a long time to bite, and give almost as much feel as Honda hydraulics.

The other changes are detail improvements — and it's nice to note that they are all things which were criticised in road tests of earlier BMW's. There are now two, clear, well placed instruments, the passenger's footrests are set further back and there is a decent ignition and lighting switch, although it's still awkward to reach in a panic. The lighting is very good, almost up the R90's full performance, a switch on the clutch prevents starting the motor in gear unless the clutch is withdrawn, and I particularly like the hydraulic steering damper. This is controlled by an offset linkage so that position 2 give a full stroke of the damper, lock to lock; position 1 gives roughly half-stroke and position 0 gives no damping.

The brakes have inspection windows so that lining wear can be seen (to comply with some law or other — all they need now is a law to make people look through the windows . . .)

After all the good things the R90 does have some faults, although you have to look pretty hard to find them. The centre stand isn't the most stable of its kind, and the fly-back propstand can only be used when you're sure the machine is securely parked on the right ground and where nobody is likely to lean against it. The front brake pads dragged enough to stop the wheel from spinning freely, and were badly affected by water.

There was also too much vibration as the motor accelerated from lowish rpm — a point I've noticed on other BMW's and possibly due to the carburettors not being exactly synchronised.

The other thing, of course, is the cost. For an average guy to buy a BMW means joining the ranks of the totally committed.

The R60 is cheaper and more economic to run, the 750 seems rather superfluous, neither fish nor fowl, for it is the R90s which steal the show as far as performance is concerned.

And by performance I don't just mean speed, I mean reliable staying power with all the luxury you would expect at such a price. In over two thousand miles of riding, riding hard enough to wear 3 mm from the tyres and wear out the brake linings, this R90 didn't need any oil, everything else stayed in adjustment and the /6 was pulling as strongly as when I picked it up. That is grand touring in the wildest possible sense.

JOHN ROBINSON

ON TRACK

Ray Knight

Test to end all tests— 24 Hours at the Bol d'Or with a BMW R90S

TAKE one R90S out of the crate a week before the world's toughest 24 Hour Race, screw everything down a bit tighter, do 250 miles and then throw it (complete with silencers and air cleaners) in with all the factory racers — madness? Stupidity even? Well there must have been quite a few sick feeling racers at the end of that gruelling 24 hours because it was running just fourteen places behind the team that cleaned three thousand cool greenies on their 1000cc fourpot double cam Kawasaki rocketship.

But to start at the beginning: Geoff Daryn had an entry for the Bol d'Or race of some weeks standing and had been holding on for some go-faster parts for a 900 engine that was laying in bits in his workshop. It was this in the first instance that was to have powered Geoff's first try at this famous French endurance test. A test everything in and on a machine — let alone the riders.

However, the best laid plans of mice and men, etc, finally time ran out and all that had materialised was a racing seat and fairing — without fittings at that. The burden of preparing what was to be our instant racer devolved upon Tony Balderstone, Geoff's car racing mechanic, ably assisted by 'Toad', who is happy to be known as just that. They took from stock a brand new bike and just did some pretty basic things, mainly they simply screwed it all down tight.

First was the changing of the clutch lining material, from fibre to a metal plate — something you could legally do even if the bike was being used for a production race, which Bol-d'Or certainly is not. Next was the fitting of Koni dampers at the rear, rearset controls, clip-ons and the seat and fairing. That's how the bike arrived at Brands Hatch the Wednesday afternoon before the race.

Tony had ridden the bike up from the shop at Sandwich, Kent, and together with one evening's running-in there was about 200 miles on the clock. The motor had been run up to 6000rpm only as yet. Brands was intended to loosen up the motor, see if we had any problems and to sort out, if required, the roadholding.

I prodded the starter button in the paddock, blipped the motor and watched the bike gyrate over to the right — and then I remembered all those stories about the dreaded torque reaction and how it was supposed to affect the roadholding as you knocked it off and laid it into a bend. That's what they said. But then 'they' say lots of things. So mentally reprogramming my feet to get the left one cogging one up and four down instead of stamping on the brake, and trying not to change gear with the brake pedal on the right side, I climbed aboard and headed off out onto the circuit to complete the running in process and to find out what BMW R90S roadsters were all about.

Buzzing off in one, two and three and finding plenty of torque in each, reactive or otherwise, I layed it over gently for Paddock Bend, hummed quietly down the hill and up to Druids where I instinctively tried to induce the right foot pedal to select second (which it does on the Trident and does not on the BMW). Thankfully the motor had enough steam to pull round the turn in third anyway and we bounded down to Bottom Bend and circumnavigated it in a series of wobbles which bode ill for a comfortable 24 hours ride. A few more laps were undertaken to see if things would improve and sure enough, I got accustomed but not too easily, so it was back to confer with Tony. We decided to wind up the damping on the Konis which were set on soft initially. We wound 'em up to threequarters damping and set the spring pre-load on the middle setting.

The improvement was remarkable and the only impediment to a really quick passage through Bottom Bend now was the thought of the slightly worn roadster tyre on the back, fitted for this session because we did not want to wear out our racing Metzeler boot. Lap times improved and the revs in the first session were gradually worked up to 6500. Then we sent Tony out to play racers and put some more miles on the clock. It was in fact the first time he'd taken a bike round a racing circuit — and he sure liked it. Hugh Evans was co-rider in the project, Hugh being of much production and 24 hour racing experience, and he went out on the bike for the third session.

By this time the revs were up to seven and as confidence was gained I mused on my merry way round Clearways that there was an awful lot of cylinder sticking out the side and that it was probably not far from the deck, but this was the first time it had come to mind. All that chat of vulnerable cylinderheads was proven to be unfounded. Going round Paddock on one occasion there was that quite unmistakable solid thump as the right head did in fact touch down as we got quicker and quicker over the bumpy surface, but it was nothing to worry about.

In the 50 or so laps that we completed during the afternoon we did find one defect in the navigationals - that we couldn't cure: when knocking the power off for Clearways when laid over, the bike had a tendency to weave — cured as soon as the power was put back on. It seemed that there was too much movement at the front and the stronger front fork fitted probably required heavier damping action to control the rebound action. We did not have alternative oils with us but Tony promised he would change it Thursday, as well as all other oils in the machine and generally check every nut and bolt.

Thursday night has us in Hugh's Transit together with the bike plus spares and wifely pit crew following Geoff's BMW 3000 (Geoff was towing the caravan that was to provide kipping space for the off duty rider). During the quick flit down to Dover we were just about keeping the back end of the caravan in sight and although Geoff had never towed before he was heaving the plot along as if in the manner born — maniac!

Crossing from Dover to Bologne and then the long drive to Le Mans, and by ten am on Friday we'd booked into the hotel and headed for the circuit to see where the action was to take place. To our delight we found ourselves placed next to the factory BMW team — they were so well organised that we even had an allotted space to park in. Imagine that at a British meeting! In fact French meetings are frustratingly well organised. We couldn't get into the race office to sign on because we hadn't a pass to get through the gate, Geoff's French was not adequate for the occasion but many gesticulations and protestations later we got officiated — it ain't very painful.

Now for the action: The bike was unloaded and it sure looked pretty with its black fairing and racing seat, and that box-stock look made it quite different from all the others with their mag alloy wheels, egli frames, open exhausts and everything else that you can think of. There was even one Japauto 960 Honda with twin calipers on each of the twin discs. That was a new one I'd not heard of but I certainly witnessed their stopping power in the race.

Hugh took the R90S out for the first practice session as I was more familiar with the track, having ridden on the Bugatti circuit twice before. He, like I, confirmed that the change to 20-50 oil in the forks had again transformed the handling and through the fast left hander out the back with its few ripples, the steering showed itself to be so exemplary the bend could be taken absolutely flat as a Manx kipper. Again let's say that it steered just like you'd expect any well set up machine to steer with no signs of any peculiarities attributable to the engine layout.

In fact when you reflect that there were two thumping great 450cc pistons whalloping across the frame at speeds up to 8000rpm, that motor was as smooth as silk.

And there was a snag. The 8000 revs that is. It was going so well that on standard gearing it was running up to eight in top and Helmut Dahne advised us that since we were intent on getting to the finishing tape firstly and foremostly, we really ought to keep the revs down to 7500. Well, we could have rolled it back, sat up, or just eased on the brakes I guess when we got to 7.5. However, bearing in mind that we were already in danger of getting run over by the opposition with their 100bhp plus aeroplanes, we felt that every mph that could be squeezed without impairing reliability was a must. So we prevailed on Helmut — a most helpful guy to get the factory to lend us one of their racing 'diffs' which would give us a higher than standard top. This had the desired effect of giving us the same top speed as the original whilst keeping the revs down nicely. After a few laps in the dark we put the bike away and went to get a badly needed night's kip — we'd missed our beauty sleep completely whilst travelling the night before.

Saturday saw Hugh losing the toss to see who was going to indulge in the mad scramble at the start and by 3pm we were lined up on the start line one hour before the race was due to start. There was an hour of preliminaries with national anthems being played while riders were introduced over the PA to the 80,000 crowd — yes, I said 80,000. There was a veritable sea of faces all round the course as shells were fired high into the air over the start area. These missiles contained little flags on parachutes, one each for each of the competing countries and as they reached 200 feet they burst open and glided earthward — and they say the Yanks are good at "ballyhoo" No chance!

As the enormous digital clock high above the start/finish ticked away the seconds tension mounted until you could cut the atmosphere with a knife. I was almost grateful for losing the toss.

At least this year the weather didn't look too bad, last year it had rained for eighteen hours and the Mead & Tompkinson BSA I'd ridden had expired. Down went the flag and Hugh made one of his usual ace starts and took the BM half way up the field from what had been a lowly grid position — decided on practice times.

Bikes crackled, screamed droned, rasped and absolutely flashed by the pits, but ours just hummed quietly on its way. We'd scheduled to run for a couple of hours at a time, trying to make up time by staying out of the pits as much as possible and at an hour 50 or so I was getting ready to go out for the first time when Hugh came in early, having had to switch on to reserve. Nearly five gallons of Total juice had been consumed.

The mad scramble with helmet and gloves took place as Tony and Toad refuelled and then I headed out for the first of my seven sessions.

Up the hill, over the crest and down to the first hairpin and one of the mighty Japautos crashed by using all those four front calipers to slow with. I reflected that I too had twin discs up front and shouldn't have lost that much time, particularly as he must have been travelling 40mph quicker when he put the brakes on. Still, I hadn't done a lap yet so made a note not to get impatient, the name of the game was not lap times — just laps.

Right hairpin, short straight, long sweeping left banked in your favour, straight and sharp right back through about 120 deg, straight then flat left, then the wriggly bit — a double "S" and back to the start/finish from a 180 deg right — I still do it in my dreams. My lap times came down and settled around 2m eight or nine seconds and I remembered Norton European Sales Manager Keith Blair saying that if his three Commandos could circulate steadily at that speed for the 24 hours then he'd be in the first 20. I had reckoned on a higher target for the R90 and had said to Geoff when pressed that given an absolutely trouble free run we could finish twelfth. There should be many of the seventy starters who would blow up or fall off.

The marshalling was really good at night. On one occasion, bounding down to the Hairpin some unfortunate had overbraked and abandoned his bike which had burst into flames — and a burning bike at night when you are racing is something you'll never forget if you ever see it. The very next time around there was no sign of the incident which had been completely cleared away within two minutes.

In this session I actually began to lap one or two machines that had been passing ours in the dry daylight, and the low centre of gravity of the BM was now contributing a lot to its handling in the wet. A few more machines fell by the wayside this session and we crept a few more places up the scoreboard to be reported as running around the 27th mark. By midnight we'd got to the lower 20s.

I came in once more and after a cup of coffee it was time to try and switch off for a while before the next spell of riding. Geoff actually foresook the pit for a while this time with the remark that the last time he'd "left his charge" so to speak — in one 500 Miler, he'd come back to find that the bike had been dropped. That remark was to be prophetic.

While I got my head down for an hour in the van some unfortunate soul had dropped it at the beginning of the start straight and deposited fuel on the road a place where you put the power on as you enter the straight, and the resulting skating rink had claimed Hugh as its victim. The rear wheel had spun on the fuel and the bike had skated up the road on its right hand cylinder head which had proved quite equal to the task of supporting it in spite of all the stories that you hear and Hugh's thoughts at the time had been "Oh dear there goes Geoff's pretty motorcycle."

In fact the only damage had been to break the throttle cable, scuff the seat and the tank was shoved off its mounting. Anyway, Hugh was handily placed for pushing straight into the pits. Everyone who finishes at Bol d'Or needs that little piece of luck and this was ours as Tony quickly replaced the tank.

The throttle cable breakage was a little more difficult as time had prevented us from making sure that the spare one we had ready would fit straight on. It didn't, but the whole affair probably cost no more than a quarter of an hour. Then there was a spot of headlamp bulb trouble to add to things and the fact that Hugh had bruised his wrist a little while I was blissfully unaware of the drama, zonked out in noddyland.

I'd started before six and by seven dusk was beginning to gather as the Munich twin hummed on its way. I'd found already where we could make time on the much faster opposition. That fast left hander demanded that the fire engines eased off to get round and we with our slower approach speed and excellent roadholding grabbed many yards back on lots of them. Up the straight and up the hill the road curves right and ripples a fair bit and the multis would welt by, almost running me over just before I got there and rolled it.

On some occasions I actually had to ease it as the road seemed very narrow. There was little point in trying to go round the outside and block the road for another quick machine that was inevitably coming up behind. I said to myself "Stay out of trouble lad, there's a long way to go."

Well it seemed that it wasn't just the dusk that was gathering as spots of rain appeared on my visor and with still half an hour of my 1h 50m to go it seemed that I was going to start the day with a right soaking. Down it came and soon the yellow flags appeared as umpteen brake horsepower fought a losing battle for grip.

In several places riders were tossing themselves up the road and already we were running about halfway up the field. It was obvious that the faster bikes were not coming by quite so often now.

I began to watch the pit intently for the welcome "IN" sign as I got steadily wetter, thinking of my oversuit, towelling scarves and rubber gloves laying in the pit while I got drowned. Then the oil flags started to appear in several places as fallen machines dumped fuel and oil in the water and produced unbelievably slippery patches. The speeds came down and down — and so did the rain.

At last the "IN" signal appeared from wife Yvonne and gratefully in I went to hand over to a fully wet-weather kitted Hugh who departed into the gathering gloom and still further slippery patches. By now one competitor had dumped the contents of his oil tank on the road for the third time. After that stint I discovered the advantages of being given the hospitality of the BMW tent which was serving hot meals to all the many back-up crews. This was because BMW Concessionaires UK had taken a friendly interest in our efforts and fixed things for us. To this we departed leaving Geoff's wife Maureen still clicking the stop watches and recording Hugh's progress and we regaled ourselves with a full four course meal and wine to boot. Dampened spirits revived rapidly.

Somewhere around ten o'clock after a short relaxez-vous in the caravan I climbed into my fluorescent pink Motomod oversuit (a special suit for racing in the rain at night), demisted my Premier's visor, pulled my housewifely Marigold rubber gloves over my racing ones and waited for Hugh to come in.

This time Tony added a pint of oil to the sump and the motor was sounding superb. To quote one of the motorcycle comics, the BMW was beginning to make its presence felt by being "unobtrusively impressive", and creeping slowly but surely up the leaderboard.

The next spell was one of easing rain, an occasional dry line here and there and heavy showers. One problem was the near dry line that developed for a while on that fast left hander. This corner is unlit and you aim up the straight from the preceding corner and wait for the headlamp beam to pick out the fluorescent marker boards — wait for the 50 yard one to come up, aim for the edge of the track just there and then heave the bike over hard and trust that you are set to clip the apex as you hurtle into the blackness. Because the fairing mounted headlamp is just that little behind in pointing the light beam where you are going, it was a shade tricky.

There was the memory of seeing Percy Tait rush by resplendent in Kawasaki green leathers, or maybe it was a matching oversuit, but hurtle he sure did. Then another DOHC four that rushed by up the straight was piloted by one of the girls that was sharing the big four and even an oversuit and leathers can't disguise that shape disappearing up the road in the headlamp. After these light interludes the "90" hummed its merry way around and around and gradually hummed up the list.

After having dropped a few places due to the spill, we retrieved several spots and were again nibbling at the lower 20s as the signals scored 30, 60 and 90 minutes and then "IN" again.

Another cup of coffee, another doze and all too quickly another call to get back to the pit. Hugh shouted that all was well and that it was drying out again before he hoofed it for the comfort of the caravan with its calor stove going full blast to thaw out with.

Just before dawn you begin to wonder what the bloody hell you are doing going round and round and round and round and then there's an incident and attention is again at full pitch as another bike slides down the road. Again fuel was dumped on the road and the marshalls with flashing torches were waving bikes down and steering them out of the way of the slippery patch and then a little while later there was the local fire brigade actually hosing down the track as I went riding by. And don't think they turned the hose away when I came along, not on your nelly. The BM and I got washed down as well.

Eventually it did nearly dry out all round and things started to speed up again. Going round that long, long left hander that I never did really master, I noticed a slight glow in the night sky and a star shinning through the clouds. Next time round the glow actually seemed to be brighter. It was dawn on the way and daylight seemed to come with a damp rush and there were a lot less bikes around now than there were about 14 or 15 hours before, but the leaders still went screaming their way along.

The Grandstands were still full of bodies but they were silent now, crashed out on the benches as the big digital clock remorsely indicated the passage of those long 24 hours. Down the box again. Top, fourth, third, second with front brake lever hard on and hands beginning to get a little weary.

Round the Hairpin for the ten thousandsth time, third, fourth, fifth, point it through the bend and the motor still seemed as if it would run for a hundred thousand miles — maybe it would at that, treated in any other way.

By now many of the pits were empty and those that were still supporting their machines out on the track were beginning to attract a following.

Faces that were strangers a couple of hours ago now became familiar. Even the Gendarme on the gate began to nod in acknowledgement of the fact that we were still racing as we passed through his point of entry (though there was still no chance that he would not want to see your pass before you went in for another spell of riding).

I completed another spell, Hugh did another also and as it neared high noon we broke the "20s" barrier and our standard racer had now outlasted the majority of the field and was running around nineteenth. It was still doing 7,500 in top up the hill and in fact seemed to be going better than ever. It was probably run in now and in fact it did not seem so slow on top speed, only coming out of the slow corners did the multis really blow us off the road.

Somewhere about this session I got involved with one of the Commandos that was still running and for nearly an hour took turns in leading or being led but then it went missing, or more likely stopped for a change of rider, and as noon came and went I mused that there was not far to go now. Surely she'd finish, I mean what's four hours after you have just done twenty? It seems such a short flit after riding all night.

Time for another baronial noshup arrived and once more I handed over to Hugh.

We were now probably 18th. Our pit began to fill up with inte-

rested supporters, some of them from the official BMW camp. It became apparent there was an awkward amount of time left to go that did not split nicely into "hour and fifty" spells so I did a shorter than usual stint, Hugh did likewise and then I took the BM out for the last session.

The pit was so full I was getting a signal of some sort on every lap. I'd never had so many in the previous 23 hours, and then with half an hour to go there was another shower and everyone who was still running slowed down so much that it was hardly true and I began overtaking people that had previously been lapping me. But everyone was intent on finishing. I had a bit of a dash round with the Negus Norton and as I passed the pit I saw Hugh had got changed into civvies and was giving a big thumbs up signal. I had time to wave back, which was probably bloody silly, but you do tend to get carried away at times like that.

Watching the other pits, I could see signs indicating the probable number of laps to go and now I can see all sorts of BMW folk in the pit and there was BMW Marketing Manager Rod Slade lending his support as I hummed by yet again. Suddenly there were people beginning to crowd the edge of the track in the final minutes. I remembered that after receiving the chequered flag it's imperitive to do one more complete lap to officially qualify as finishers, and after all that preceding 1,500 miles, nothin' was going to stop me — if I had to carry the bloody thing round on my back.

Next time round the track was completely blocked by spectators all shouting and waving. There was a man somewhere up there with a chequered flag and marshalls frantically trying to keep a couple of feet of track clear to let the bikes go through the flag for the last time. I had to run over a few French toes as I got thumped enthusiastically on the back and cheered and waved and pushed through the narrow gap.

The motor was so quiet that the crowd didn't hear the bike and I had to push through them, revving the motor hard. The crowd was just damn crazy.

I set out on that last lap in convoy with a couple of other survivors of that incredible marathon. The marshalls on the corners came onto the tracks and cheer us as we dribbled round in second gear. Suddenly it's all over.

Hugh took the bike from me — we have finished fifteenth. Rod Slade came up with a bottle of champagne and I got a quick soaking in champers as we all grinned like it was one great big joke.

We'd completed what must be surely the toughest racing test for a standard bike that had ever been devised. 250 miles out of the crate, a few suspension mods and then belt hell out of the thing for 24 solid hours.

My thanks go to Tony and Toad for preparing the bike, Geoff for backing the project, Eunice, Yvonne and Maureen for standing out in all that rain all that time, Hugh for keeping the seat warm while I was away and to that big black beautiful Munich twin which I have just developed a very healthy respect for.

As a sort of postscript; I reckon that in a production race, as opposed to the sort of lunacy that we were involved in, that it would be more than competitive and I intend finding out just how it will go in its own class — next year now. Now I've just got to have one for 1975!

JUST PASSIN' THROUGH...

Chasin' The Leaves On An R75/6 BMW

By D. Randy Riggs

THE IDEA HAD flashed through my mind late last year as I drove to the funeral of a very special lady in the East. "Brownie" was her name, and the day was one of my saddest. But I knew she'd rather have me thinking happy thoughts than miserable ones, and that's when it hit me. It was November, but one of those very special days when the sun is bright and the air is warm. Indian summer in the northern New Jersey mountains got me to thinking what a perfect day it was for a motorcycle ride. And having been born and raised in the East, California hadn't begun to wipe out the memories of the seasonal changes; the weather mixes that are non-existent in the Coastal Southwest. I was too late for the best part of fall, but I knew then that October would come again and I would be there to greet it. On a motorcycle.

The thought never really left my mind all year, and when the month drew near I contacted Butler & Smith, the people who import the BMW into the U.S. Would it be possible for me to pick up a 1975 model R75/6 at the Norwood, N.J., headquarters for a touring trip of the Northeast? "Yes it would," they said. "And, oh, any objections to having us fit it with our new MK-VI fairing and Krauser saddlebags?" Heck, you took the words right out of my mouth.

Since I wanted to hit the area when Ma Nature painted the leaves with the brightest colors on her palette, timing would be critical. When the leaves decide to change, they do so quickly. The best period might only last a week. I kept calling friends and relatives back there for the latest updates on the conditions, finally heading East the day after the Ontario road race.

Flying into Philadelphia I kept peering out at the landscape as the DC-8 made its approach. Was I too late?

It didn't look it.

At Norwood the 750 sat waiting. Black too. My favorite color on a BMW. But I'm not alone; about 60 percent of the BMW models sold in the U.S. are delivered in Henry Ford's favorite shade. And this was a 1975 model, so I was anxious to see what had been "improved." Things don't seem to "change" from year to year on the BeeEms, just "improve."

Right off I noticed new handlebar switches and different seat upholstery; flat black turn indicators too. "What else?" I said. They pointed out the remainder of the differences like "click" action fuel petcocks (introduced in mid-'74) a drilled disc front brake, new brake and clutch levers, ignition switch with "marked" positions and the lack of a kickstart mechanism. "Anything else?"

"Aw, just a bigger starter motor and stronger electrics."

"Sounds just fine. See ya in a fortnight. *Ciao*...er...I mean *auf Wiedersehen*...heh, heh...bye."

I went outside to load the machine. Though I travel as light as possible, it seemed as though I had too much to carry for my intended two-week excursion. I was concerned whether I'd have enough room, but I shouldn't have given it a second thought. The Krauser "Deluxe" Number 272 saddlebags swallowed everything up and had room for more.

The Krauser bags are the absolute best I've ever used, but, for $265, they should be. Included with the bags are chrome brackets and a luggage rack that bolt to the machine in minutes. The bags themselves snap on and off in a second with a push-button release mechanism. They're waterproof, lightweight, look like they belong on the bike and can be used as regular luggage. The only bad feature I could find was the lack of a lock to secure them to the machine. As a result I always parked where the BeeEm would be in my sight. Loaded to full capacity they hold 22 lb. each and there's even enough room in one for a full coverage helmet. Look for the Krauser bags at any BMW dealer. They're the answer.

Worrying about packing special tools is the last thing a BMW owner ever does >

R75/6 BMW

and I didn't either. Underneath the lockable, flip-up seat sits a removable tray containing a full complement of high quality tools; if you can't fix it with any of these, then you couldn't fix it at home in your own garage.

Along with a particularly good owner's manual, there's a service directory for BMW dealers all over the world, tire pump, patch kit and enough wrenches to turn every nut and bolt on the entire machine. And, should you get your hands dirty, there's even a rag emblazoned with the letters, BMW, to wipe your hands on. What more could a rider ask for? Even the centerstand allows the front or rear wheel to be removed with no outside help in case of a flat or something equally awful.

The weather was perfect that particular day, but it was getting late and I didn't want to be riding much past dark. In October an Eastern night can get downright cold, and I didn't want to ride past all that dynamite scenery in the black of night. I stuffed my maps of the Northeast states into one of the fairing's storage compartments and checked out the new handlebar switches.

Appearance wise, I think BMW has hit the nail on the head. These are without a doubt the most unobtrusive and attractive control devices I've ever seen on a motorcycle. The left side controls horn and lighting. The horn (a loud one), operates with the touch of the upper left rocker switch; the lower switch (rocker type) controls high and low beam lighting, as well as the high beam flasher for saying "hi" to other BMW riders. A day-glo colored switch turns the lights to "parking" or "running" position. All operate easily within reach of the thumb. Terrific.

On the right side, the layout is identical, but the day-glo switch is an emergency kill button, and the rocker switches operate the starter motor and turn indicators. All are clearly labeled, but there may be one flaw and that lies with the turn indicator switch.

The switch operates in an up or down manner, rather than side to side. So the rider has to memorize and get used to "up" meaning a left turn and "down" meaning a right turn. In actual practice it's confusing until it seats in your memory.

With the choke full-on, the 750 fired instantly; as I let it warm up some, I noticed that it had only been ridden four miles. "Whoops, this thing's brand new...better find out about break-in." They almost laughed when I asked. "Just run it and don't worry." Nothing to worry about seemed to be the story of this motorcycle's life. So I didn't. And I rode away.

Smells of fall were in the air that day as I motored through small towns that'd seen many more seasons than I. Work day traffic crawled slowly in places; most of the vehicles' occupants eyed the BeeEm with curiosity. Maybe it was the California plates, perhaps just the gleaming black beauty of the new machine. But I'm sure there was some envy for anybody riding a motorcycle on such an extraordinary afternoon. Even a milktoast would have forsaken his car.

I broke loose from the 5:30 traffic when a few of the towns thinned out and many of the commuters reached their destinations and after-work gin and tonics. Big yellow school buses trundled by on occasion, lightly loaded with tired athletes fresh from their workouts. Once in awhile I could smell good things cooking for dinner; an eager paper boy tried to pace the BeeEm and I on his MX-style bicycle, but I lost him when his concentration faltered and he had to aim an Evening Telegram up someone's driveway. He skidded out of sight in my rear view mirror.

Big orange pumpkins were turning up on porches here and there. Stores had specials on Halloween candy. Grotesque masks hung in five and dime windows. It wasn't hard to tell that summer was over.

After getting lost once or twice, I finally got headed in the right direction and arrived at the night's destination. Kittatinny Lake...late. It was very cold and crisp, and all my warm clothing was packed in the bags, so I was also cold and crisp. My favorite aunt and uncle looked surprised when I removed my helmet; it was their house where I'd be staying for a few days.

"You didn't have all that hair and the beard the last time I saw you," my aunt said, still looking somewhat shocked.

"And I wasn't this cold," I replied, trying to bend my legs into a walking position.

When you mention the state of New Jersey to a Californian who has never seen the place, or seen only Hoboken from a turnpike bridge, he is very unenthusiastic about it. "Why on earth would you want to go *there*?" is the normal reaction. But these geeks have never learned that Jersey got its nickname, "The Garden State," for a reason. The next day, after a good night's rest and thaw-out, I did some exploring through some of the prettiest parts of northwestern Sussex County, in perfect fall weather.

The BMW had no objections whatsoever to the winding, sometimes bumpy roads of the Kittatinny Mountains. The front forks would just "whoosh" up the bad holes and glass over the little ones, like a baker smoothing the icing on a birthday cake. Some people criticize the softness, but on a touring machine it's the ultimate. Some of the sharp dips in the road were fun to run through fast, feeling the G-forces turn the pit of my stomach hollow for just an instant.

Only a few weeks earlier I had spent some time on an R90/6, but the 750 felt much smoother. It hummed along quietly, seeming to enjoy the ride almost as much as I was. I stopped for a bit by the Dingman's Ferry Bridge crossing the Delaware river, and I listened as the cars rattled the planks of the bridge's floor. They sure don't build bridges like that anymore.

The BMW rattled the planks, too, and on the Pennsylvania side of the structure there's an old toll building with a roof covering the road. There wasn't a troll in sight. A lady walked out as I approached and readied her change pouch for my quarter. "You've picked a lovely day for a ride," she greeted me.

"You said it," I answered, as I ka-klunked the BeeEm back into first and pulled away, feeling instantly superb. It was one of the most pleasant feelings I had had in some time, because everything was just...well, *right*. It made me think of a song by Elton John with the exact title for what I was feeling then..."Harmony", and the words...

Harmony and me
We're pretty good company,
Looking for an island
In our boat upon the sea.
Harmony, gee I really love you
And I want to love you forever
And dream of the never, never, never
leaving harmony...the words were just perfect.

Autumn colors were at their peak; my timing had been lucky. In the Wallpac Valley I noticed many homes deserted and boarded up...or broken up. There must have been a reason for it, so I asked. The government and its infamous Army Corps of Engineers had decided several years ago that they wanted to build a dam and flood the Delaware river into the Valley. For their usual reasons. But, thank goodness, the public started giving them lots of flak about the project and it doesn't look favorable at this point for the dam. But at the same time the government decided to make a large portion of the area a recreational zone, so many of the people had to move out just the same. It was depressing to think about, but then again, it would keep out land developers and real estaters. Those are the guys who could turn the world into a wall-to-wall subdivision overnight if you let 'em.

In a few days I had to move on to points north, so it meant saying goodbye to several new friends...and to the perfect weather I had been enjoying. They wished me well, and offered warnings about the ticket happy troopers turned tax collectors enforcing the 55-mph speed prohibition.

I tried to go 55 for a bit, but on two-lane roads I was holding up traffic and on thruways most cars and trucks were whizzing past 20 mph faster. So I joined 'em. Even the 750 seemed happier. I sure was.

The hardest thing about touring in the East is that there is so much to see in a compact area. I could spend a month exploring the towns nestled along the Delaware River's edge and not begin to see it all. Towns like New Hope, Washington's Crossing, Lumberville, Yardley, Milford; every one full of history. But I was already into upstate New York, nearing Middletown on the Interstate, heading for Providence, R.I.

Seating position on the 750 was just right with the fairing in place, fairly straight up and erect. But take that fairing off and I'd much rather have the bars that come on the R90 Sport in order to stay down out of the wind. I had jacked the rear shocks up to their stiffest position back at Butler & Smith, and now at 70 mph I was playing with the three-position steering damper, except that I couldn't seem to rotate the knob. "Strange," I thought, and I pulled into a rest area to have a look.

It was an easy fix. One of the hose clamps used to hold the MK-VI Butler & Smith fairing in place had been positioned to interfere with the operation and arc of the damper. I loosened the clamp and rotated it out of the way; then the damper worked fine. I kept it in the number two position on the higher speed roads.

Sidewinds were very strong as I crossed the Hudson River atop a spectacular bridge near Peekskill, my run through Connecticut was short and to the point. It was beginning to rain. When my throttle hand got tired of keeping a twist on the grip, I screwed down the throttle stop screw and let it do the work.

The rain stopped, but it remained dismal. In some areas the leaves had not yet turned; in others they were long past their brightest stages.

I was able to travel 170 to 190 miles on the 4.75-gal. fuel tank before it went on reserve; the 750 was averaging more than 50 mpg. Somewhere in Rhode Island I stopped to stretch my muscles, and I was surprised to notice that the rear wheel was getting messy with oil. The drain plug on the rear drive was dripping; it lost a teaspoonful as I sat there and watched. I tried to tighten the plug but it was already plenty snug enough. "Hmmm, the first BMW I've ever seen leak a drop of oil." Some

Photography: Gene Dwiggins, D. Randy Riggs

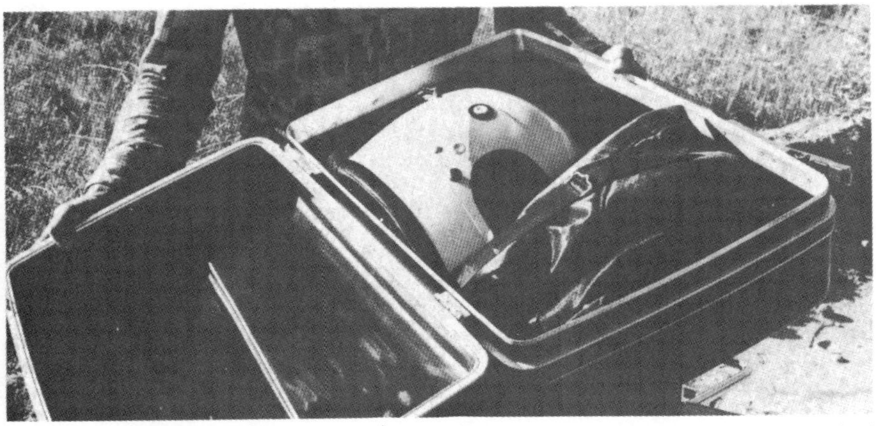

gasket compound later cured the problem.

Before I headed to Maine I spent a few days with photographer Gene Dwiggins and his wife Margie in the city of Providence. They belong to the Rhody Rovers MC, and it was interesting to hear how off-road riding is in this part of the U.S. And Margie cooks good too.

I don't think they were anxious to send me off into the pouring wetness, but I really had to get going and I prepared for the worst.

I snapped on the Vetter "Hippo Hands" I had brought along, jumped into everything warm I had and covered it all with my Full-Bore riding suit. "Let's go BMW...and see how you like the rain."

It was awful, but I was staying dry. The fairing kept much of the blast of rain off, but it was coming down so hard that some cars had even pulled off the side of the road. My face shield would fog up every few minutes, so I tried keeping it open about an inch. That cured the fogging, but the rain dripping off my nose tickled. Which was the lesser of two evils I never decided.

I loved watching people in cars staring at me; I had kept it to 55 in the rain, but they kept zooming past. I thought, "Bet I'm the only person in Massachusetts on a motorcycle right now." I was wrong. A few minutes later I heard this vroooom sound and I looked over and here was this dude on a Moto Guzzi, with a sidecar no less, looking for all the world like he was really digging on the rain. We waved very courageous waves, like, we were the only two nuts with enough guts to be out in this stuff...like being brothers. He eventually peeled down an off-ramp and that was the last motorcyclist I was to see for some time.

When I got to Maine it was still raining and cold, but the leaves were pretty in spite of it. I got on the Maine Turnpike and a sign greeted me. "Welcome To Maine," it said. "Speed Limits Strictly Enforced. Patrolled By Marked And Unmarked Cars, Aircraft, Radar In Use. Your Cooperation Is Appreciated." Thanks Mr. Big Brother, for the warm welcome. I was intimidated for a few miles, until the trucks rolled by at 80. I screwed it on too. To hell with 1984.

At Augusta I picked up a two-lane road again and headed for the coast. When I got to Belfast the rain had finally stopped, but I was only a few miles from Searsport, where I had planned to stay and see some old friends.

Maine is a state like no other, very busy with tourists in the summer, but seemingly deserted in the off-season. My old friend Harry Gray used to tell me in his wonderful Maine accent, "We've only got two seasons up here...July and Winter." And he's right. I spent some time with Harry the following day, and he told me about the motorcycles he used to ride...a Pope, Henderson, Harley, a couple of Indians, and he told me about the time his wife Jessie sold his last Harley when he was out to sea on an oil tanker. "I was sure fit to be tied when I found that out."

We bid our goodbyes and he told me, "If I was a few years younger I'd be going with you." I knew he wasn't kidding.

PARTS PRICING

Warranty	6 mo./6000 mi.
Air Filter Element	$7.50
Rear Tire (standard)	34.97
Headlight Bulb	3.25
Taillight Bulb	.50
Turn Indicator Bulbs	.50
Battery	35.00
Clutch Cables	5.55
Throttle Cables, ea.	(2) 6.00
Brake Cables	5.45
Ignition Points	
Points	3.95
Condenser	3.10
High Tension Coil	19.85

A cold front had moved into the New England states and though it was clear and sunny, it was bitter cold and windy. Dry leaves scampered about in the rushes of the breeze, and the limbs creaked in the trees, as if to tell us winter was on its way. The frost had come sharp and quick as driven nails, the leaves had turned their blazing reds, and their bright yellows...and now many were dry and brown and dead. Autumn can be a sad time, and you know it's autumn because the hunters are walking in the woods and fields, their hounds barking their presence; cider presses are getting a workout, the corn is shocked, fields are mown, the earth is waiting, preparation for winter goes on in earnest. October is the season for returning...and I have. But now it's time to head the BMW back south down the coast...my trip was getting short. My stomach growled, so I turned into the parking lot of a cozy-looking diner.

A stack of Republican Journals sat at one end of the counter and the music box had an interesting blend of rock and country choices. But the waitress was into Elton John and that's what she kept playing over and over with the quarters she'd earned that morning. I didn't mind. She asked me about the BMW and said she wished she could take a ride like that someday. "A lot of people tell me that," I said.

The meal was great and so was the coffee, but as I was about to get up and pay the check, the little dear played "Harmony." She filled my cup one last time as I listened, and I was in the perfect mood for the final leg of my trip.

Down US 1 the BeeEm purred, through the breeze off the Penobscot Bay, coastal Maine smiling back at us. I crossed a bridge over what the sign told me was the Great Brook, and it made me smile. Towns are just the way they appear in all the pictures you've seen and being "Down East" is a perfect way to use up a ride.

It got colder, and I stopped in the parking lot of "Bernard Nelson, Useful Cars," to slip on my warmest gloves. The end of the trip was fast approaching, almost time to stop all this gazing and smelling and listening, and almost time to give up this wonderful motorcycle. I followed all the signs that read, "South" and, in time, it was over.

But it never really hit me until I was aboard the jet, watching Philadelphia, Baltimore and Washington slip under our wing, seeing the lights of the Eastern Seaboard blend into the darkness behind us. That's when I was sad...because it was over.

R75/6 BMW

SPECIFICATIONS

List price	$3475 (as equipped)
Suspension, front	telescopic fork
Suspension, rear	swinging arm
Tire, front	3.25-19
Tire, rear	4.00-18
Brake, front, eff. dia x width, in.	10.24 x 1.52
Brake, rear, diameter x width, in.	7.87 x 1.18
Total brake swept area, sq. in.	73.87
Brake loading, lb./sq. in. (160-lb. rider)	9.39
Engine, type	ohv opposed Twin
Bore x stroke, in., mm	3.20 x 2.78, 82 x 70.6
Piston displacement, cu. in., cc	45.4, 745
Compression ratio	9.0:1
Claimed bhp @ rpm	57 @ 6400
Claimed torque @ rpm, lb.-ft.	43 @ 5000
Carburetion	(2) 32mm Bing constant velocity
Ignition	coil and battery
Oil system	gear pump, wet sump
Oil capacity, pt.	4.75
Fuel capacity, U.S. gal.	4.8
Recommended fuel	premium
Starting system	electric
Lighting system	12V alternator
Air filtration	dry treated paper
Clutch	single-plate, dry
Primary drive	none
Final drive	shaft and bevel gear
Gear ratios, overall: 1	
5th	4.64
4th	5.16
3rd	6.40
2nd	8.84
1st	13.60
Wheelbase, in.	57.7
Seat height, in.	32.0
Seat width, in.	12.0
Handlebar width, in.	28.5
Footpeg height, in.	12.0
Ground clearance, in.	7.0
Curb weight (as equipped) (w/half-tank fuel), lb.	534
Weight bias, front/rear, percent	43/57
Test weight (fuel and rider), lb.	6.69
Mileage at completion of test	1550

TEST CONDITIONS

Air temperature, degrees F	62
Humidity, percent	41
Barometric pressure, in. hg.	29.99
Altitude above mean sea level, ft.	383
Wind velocity, mph	4-6
Strip alignment, relative wind:	

PERFORMANCE

Top speed (actual @ 7020 rpm), mph	107
Computed top speed in gears (@ 7200 rpm), mph	
5th	118
4th	106
3rd	85
2nd	62
1st	40
Mph/1000 rpm, top gear	16.4
Engine revolutions/mile, top gear	3679
Piston speed (@ 7200 rpm), ft./min.	3325
Lb./hp (160-lb. rider)	12.17
Fuel consumption, mpg	52
Speedometer error:	
50 mph indicated, actually	46
60 mph indicated, actually	54
70 mph indicated, actually	64
Braking distance:	
from 30 mph, ft.	36
from 60 mph, ft.	135
Acceleration, zero to:	
30 mph, sec.	3.1
40 mph, sec.	4.0
50 mph, sec.	4.9
60 mph, sec.	6.0
70 mph, sec.	7.4
80 mph, sec.	9.2
90 mph, sec.	11.8
100 mph, sec.	15.0
Standing one-eight mile, sec.	9.80
terminal speed, mph	82.19
Standing one-quarter mile, sec.	13.95
terminal speed, mph	94.26

ACCELERATION / ENGINE AND ROAD SPEEDS / RPM X 100

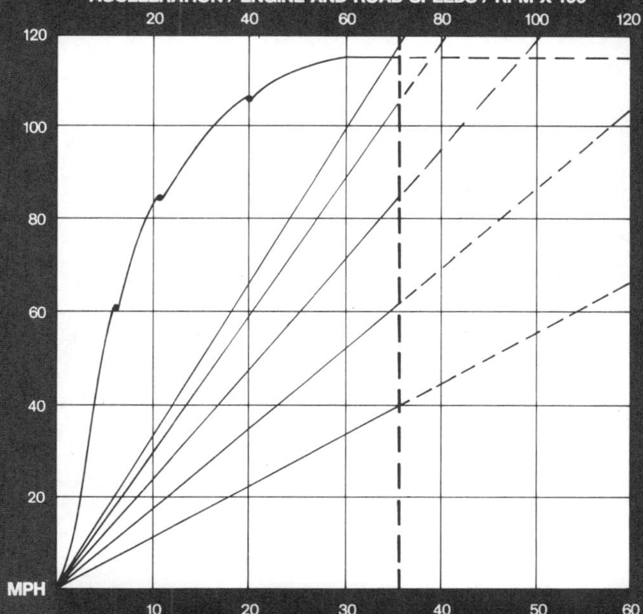

Short of food, fuel and water . . . but never sand.

By Daniel Richard

NOBODY BELIEVED it when I told them. "You're going to what?" they gasped. "Cross the Sahara Desert on a motorcycle? But it can't be done, can it?" As though it could, I snapped my fingers and mumbled something about making pies. But could a motorcycle even make it across the Sahara?

In early September my brother and I rendezvoused with my old Peace Corps buddy and his Swedish girl friend in London. Fran and Shasti shared a silver BMW 750, while Randy's classic black BeeEm differed only in color from my white R60/5. We broke them in gently on England's country lanes, and headed south toward North Africa. In mid-December when we cruised into Algiers, Algeria, I looked up an old-time oil prospector and pumped him for every drop of his desert lore. "Take along plenty of gas," he advised, "Eat dates for your energy and onions for your water. You'll be all right." Then he told me how he had walked away once from his Land Rover, and was lost within 80 yards. "The desert is tricky," he warned, "but keep your rig cool and use your head." Thanks. He whistled when I told him I was riding a BMW 600.

Even on the Michelin map our route looked impressive. It ran 2100 miles from Algiers on the Mediterranean, down the length of Algeria and Niger to Kano, Nigeria. A long haul, especially over the 1300 miles of washboard and sand between the oases of In Saleh, Algeria and the town at the tarmac called Zinder in Niger. On this section lay "The Hump"—a great sand sea—straddling 560 miles from Tamanrasset to Agadez over the barren no-man's-land separating Northern Arab Africa from sub-Sahara Black Africa.

To cross this region the British Automobile Association said, the *gendarmes* gave authorization after a close vehicle inspection for desert worthiness at the last oasis of In Saleh. The AA also wrote: "motorcycles with or without sidecar would probably be prohibited." Five days after New Year's 1974, three BMWs approached In Saleh not knowing if their dream was at the beginning or the end.

The only paved street in In Saleh is around the gas station, where spilled oil has asphalted the sand. Fish-tailing in the beach-like street, we pulled in front of the police *Daira*, and jockeyed around for a hard place to stand the bikes.

As we approached the verandah, I readied myself for the cross-examination with a couple of French sentences about the infallibility of BMWs and about three being a virtual convoy. An officer was waiting for us.

When we handed back the completed forms identifying ourselves and next of kin, he looked slightly skeptical as he checked our answers. On the far wall of the office a sagging shelf balanced what looked like the bundled results of a Boy Scout paper drive. Ink, stamp. Ink, stamp. "*Voici, Monsieurs,*" he smiled and handed out our authorizations to travel.

"*Merci, Monsieur!,*" we all said in unison. "*Merci, beaucoup!*" In the French style we traded handshakes all around and our benefactor joined us to admire the bikes. There was never a hint of inspection, but then there would not be the hint of a search after seven days if we never checked in at Tamanrasset.

Attempting jovial exchanges in our broken French, we all shook hands again.

Subduing The Sahara

"*Bon courage*," he called as we started out of the office.
"*Merci.*"

Bon courage, that is the call of the desert. Not *bon voyage* or "have a nice trip," but *bon courage!* If the going gets rough, you're on your own, so take heart, hang in there and it will get better up ahead. The call of the Sahara: *bon courage.*

We pitched camp in the dunes near In Saleh, pushed our stakes into crusty sand, and turned our last attentions to the bikes. An hour after dark, freezing temperatures would drive us into our down bags. In fact, during the winter on the Sahara, you are more likely to freeze to death than die of thirst. Heavy with gas, water and food, we had trimmed our spares down to the essentials: points, plugs, cables, tubes, condensers. Fresh oil and air filters armed us against the dust on the *piste*, and kidney belts proved as necessary as a helmet. *Piste* is the French word for "track" which is pronounced just how it makes you feel.

From a distance we looked like a tanker convoy. Each bike hefted almost 20 gallons of liquid to carry us to Tamanrasset—the only reliable gas station on the thousand miles between In Saleh and Agadez. In the end we did not stock in onions, but, instead, half a gallon of water each per day for 10 days. We estimated our gas mileage at two-thirds the normal tarmac average, adding a 25 percent reserve. Hopefully our advisors knew more about gas consumption than they did about *gendarme* inspections.

That night the desert was silent like only the desert can be—not a leaf to rustle, not an insect chirping, not a sound of man for miles—only the spectacular stars filling an infinite dome from horizon to horizon. In the Sahara silence we slept like mummies.

The first day out the bikes were heavy and sluggish under the load. Feeling fortunate not to have a passenger, I consoled my shocks by thinking every mile made it lighter. The others were up ahead as I rode into a bend, concentrating on the outside sand rut. At the last moment I lifted my eyes only to spot two Citroens coming down the same rut. Slide, plop, I nose-dived into the sand, pinned to the *piste*. Seven hundred pounds of bike and gear lay on my leg.

As my face bubble gritted in the sand, I felt like a football center after a quarterback sneak. . .trapped. Helpless, I set my jaw against the cold feeling of petrol soaking through two pairs of Levis, when I heard a car door slam. The Swiss guys in the Citroens ran over and lifted off my mechanical fullback. Nothing hurt. We traded stories—one of the travelers had his hand in a sling—told each other what kind of road to expect, then with thanks again, I kept on trucking south.

Three days beyond In Saleh, the *piste* came upon a Muslim *marrobot* shrine, which are usually consecrated by the burial of a holy man. As we watched, two giant tractor-trailers made a tight circle around the little white-washed shrine. The driver said that paying respect to the Saint guaranteed a safe journey. Needing everything we could get, we joined the faith. Once around to be safe, then three times to seal the deal.

Bang on with grace, we ticked the clicks toward Tamanrasset. Kilometer after kilometer the markers bounced by: Tam 230, Tam 225, Tam 220, Tam 215. Someday somebody would ask me, "You ever been to a place called Tamanrasset?" And I would say, "Tam? Sure, I been there. That was where the real desert began."

Anyone who is going places in

Near Agadez, Niger, in the thorn scrub of the Sahel region, a herd of camels passed unconcerned as we broke camp. Though we all carried a spare rear tire, none of the bikes had a single flat on the desert—probably a record.

Tamanrasset meets at the gas pumps. The first afternoon there we buttonholed truckers and travelers hot off the southern *piste* for their reports. Sand, sand, lots of sand was all we heard. No one encouraged us to attempt "The Hump." That night Fran and Shasti decided not to risk two on a bike and arranged for Shasti to ride to Agadez with five Canadians we had met at the *marrobot*. Tomorrow—after customs and immigration—the desert grapevine would know us as "the three Americans on motorcycles" like a newspaper headline.

Next morning, 50 miles south of Tam, the Hoggar Mountains fell behind us, and we throttled onto a great sand sea as flat as the ocean and giving the same illusion to tempt a survivor to "go have a look around." The sun climbed on our left, and with it came the wind.

The heat rose and the sand-ladened *sirocco* wind quickly turned into a stinging blast obliterating any trace of the track. Deciding it was foolish to guess, we parked the machines' tails into the wind and stopped the pipes with rags. There was nothing to do but wait out the storm. We stood around like bulky astronauts, wearing our goggles and helmets and gloves, and grew to appreciate the usefulness of the Tuerreg's flowing turban called a *cheche*. By noon the sand storm let up enough to expose the track.

Two hours later we were approaching the border where ragged children began to race across the flats trying to make our paths meet. One teenaged boy made it. After we gawked at each other, admiring his sword and my bike, he made a gesture like tossing a handful of M&Ms into his mouth.

"Sorry, no candy, but have a cigarette."

He accepted them in cupped hands, put two away for later, then expertly emptied the tobacco of the third into his palm and popped it all into his mouth. Why waste a precious match when you can chew?

Dusk stopped us that night after crossing the Algerian border at In Guezzam. Tomorrow would mean Assamaka, and I remembered the Dutchman's words in Tamanrasset: "Transport lorries stuck in the sand there, much sand."

He was right. From a distance, shimmering in the heat haze, Assamaka was two palm trees and a pueblo of low buildings in the middle of nowhere. There was no road, only a line of posts leading to the border station. Struggling through the sand mile after mile in first gear made us recheck our mileage while we waited for our passports at the oasis. Bad news. Besides using more gas feathering the clutch in low gear, evaporation sucked our supply, and the average mileage fell below our estimate, eating into our reserve. Luckily, a few gallons from a Canadian couple in a zebra painted camper did not jeopardize their supply. At one o'clock, when normal nomads napped, we hit the *piste*. It was our biggest mistake.

A quarter of a mile out of sight of Assamaka, we discovered what the Tuerregs call *feche feche:* soft, churned, bottomless sand. We were once told the desert Tuerregs have as many names for sand conditions as Eskimos have for snow. If you kept speed, your front wheel soon jackknifed in the mounds. If you slowed, your rear wheel sank into the sand like a hot ball bearing into Crisco. We tried everything. After each try we could smell our clutch plates. Our tires smoked in the sand. And the sun made metal too hot to touch. The desert had come to get us.

In one horrendous effort, the three of us galloped and cussed one bike through a 30-yard patch.

During the winter on the Sahara, you are more likely to freeze to death than die of thirst. At this campsite on the Tademait Plateau between the oases of El Golea and In Saleh, frost can be seen on the seat of the middle BMW.

The only luxury in the Sahara is the scenery. This is a view looking down from the southern escarpment of the Tademait Plateau toward the well at Ain el Hadjadj, halfway between El Golea and In Saleh.

"Hey, man, our clutches won't take much more of this," Fran said.

"Christ, it's hot," Randy unzipped his Barber jacket. "They said it was 121 at the oasis."

We passed around the squeeze bottle of water; just as refreshing and the spray used far less water. Without speaking we trudged back to bring up the other machines. My brother picked a good path and pulled away from us.

"Ride it, Ran!"

"Goose it, baby! Goose it!"

Then the bike faltered, slowed and settled down like a mother hen on a nest. He stepped off and the BMW just stood there up to its pipes in sand. The glare of the desert made us squint.

"Damn, thought I had it." Ten feet farther was the end of the patch of *feche feche.* Frustration turned on us.

"What if it's like this all the way?"

"What if it isn't?" snapped Fran.

"You can't go far without a clutch," I countered. "And in this heat we'll lose it right quick."

"Yeah, well maybe you want to stand here until the Boy Scouts come to save you."

"Gent bent. What I said was, it's the hottest time of the day. Let's stay cool, take a break and try later when we don't have to fight this heat too."

"Hell, man, if we quit each time it gets tough, we'll never make it. What do you say, Ran?"

"It may get better up there," he nodded toward the next marker post. "I think we should give it one more go."

"Okay, one more go," I conceded. I wished Fran would burn out his clutch just to prove he was not always right.

Fran went first as usual. He picked up momentum, steadied his line in a tire track, plowing through the soft sand. As he sank, we ran behind him lifting his back end to equalize the drag of the sand. Sand sprayed from his wheel like a water skier's fantail. Unexpectedly, the desert hardened and he galloped away from us leaving a trail like a waddling turtle. Randy was next.

Taking an education from the first track, my brother rode his "African Queen" hard and strong up next to Fran. Bogie would have been proud. Following suit, I shot on past and we all began to cruise again over the gravel flats. The breeze cooled our nerves as we put the sands of Assamaka behind us.

It had been 16 days since we left Algiers. Riding day after day due south, 1500 miles into the Sahara. Few vehicles traveled "The Hump" and only the occasional camel greeted us between the markers. Southward, always south. Suddenly we all pulled up short. We were lost.

"Did you see the last post?"

"No, I was following you. Last one I saw was before all those camels. Can you spot the next one?"

"I thought that was a marker," Fran said pointing at a scraggy tree. "Everything wiggles in these heat mirages."

Randy looked up from the map, "According to this, we go straight. I wonder if those camels back there were at a well. Abangarit should be around here somewhere."

Then for the first time we noticed we were surrounded by hundreds of thousands of camel footprints.

"That Englishman's instructions say we should turn a sharp left at the village of Abangarit near a ruined house. But he's been wrong before." After weeks of riding south it seemed strange to me to suddenly change direction.

"Do you suppose that well was Abangarit?" Randy remembered, "There was a pile of bricks there."

The only thing to do was follow our tracks back until we recognized some-

Continued on page 141

Fran Mola takes a dusty friend for a once-in-a-lifetime test ride on a BMW 750 near Agadez, Niger. To protect against flying stones, we all covered our headlights.

R90S

John Robinson samples the loneliness of the long-distance commuter

At the top of BMW's line the R90S is the most powerful, most luxurious and most expensive model. The only annoying thing about the 90 is that BMW could have made it several years ago. As to the cost, a garage man summed that up while he was filling the tank with about four pounds worth of petrol.

"Isn't that the £1900 one? And these are the ones you mustn't fall off, arên't they? God's teeth, £1900 for a motor bike."

And he was near enough right, at least for a few more pounds you could buy the cheapest BMW car, not that anybody's ever accused them of being cheap and nasty. I still can't figure out why a two-wheeler should cost as much as a fairly luxurious car; even using a lot of pressed steel, bigger quantities and more mass production all round. Henry Ford did it around 1920 and promptly shut down four-fifths of the American motorcycle industry. Would BMW do that to themselves?

Probably not. You only buy a motorcycle like the R90 precisely because you do not want a car. So do they make a vast profit?. Are the import tariffs that high? What do you get for your money? What exactly that no-one else supplies? About the only items that aren't to be found elsewhere are a clock, a voltmeter, disc brakes with holes

in them and a prop-stand that shuts itself.

You also get the more intangible qualities which would need to be measured in units of personal effort per mile per year, if that were possible. And the reliability, longevity plus a high trade-in price make it a sort of investment which is just the point BMW use to sell their machines to police forces. After 50,000 miles, after 100,000 miles, you've still got a BMW.

In the last two or three years BMW have changed all of the points which we, and other magazines, had criticised and there isn't a lot left to complain about. They've got their flat twin to the point where any significant improvement would be accompanied by equally significant disadvantages, if only in terms of cost. The engine doesn't need any more power, you hardly ever have to use all the power available as it is and any extra would cost more in fuel consumption and probably in reliability too. The handling, considering the level of comfort, is better than you get anywhere else. I know that BM's sometimes suffer instability, particularly when they are carrying a tail load, and this is something which should have been sorted out a long time ago. But to make a machine which is measurably better all-round, BMW will have to make a completely new motorcycle. There would be a lot to be said for a machine which provides good weather protection, for example, and I suspect that there's more to this than just slapping a fairing on an existing motorcycle. And nobody could claim that an engine which only produces 30 per cent of the potential power is anywhere near perfect.

Even so, 45-50 mpg with 90 mph cruising and bursts over 100 mph is as good as you can get anywhere else. Lots of bikes will go equally rapidly but none so easily and comfortably that you can ride for several hours and still look forward to the next stint. Actually there's not much point in running a BMW unless you can get this much out of it. It's a waste of everything to merely travel ten miles to work every day; like having a mansion and only living in the east wing.

The performance of the R90S is being able to run down to under 1000 rpm in all the gears; is being able to reach 100 mph in a quarter of a mile and being able to stop in less than half that distance; is cruising at seventy and still averaging sixty; and

s travelling over 200 miles 'ore stopping for petrol.
There are bad points too. jearbox which clunks when ı change gear below about 00 rpm; a remarkably ooth ride which still suf- s from blurred images in mirrors; and having to e the tank off to check the ke fluid level.
The R90S has different burettors to the 90/6, they l cause vibration when y get slightly out of synch these types seem to w a little more power at top end and to let the jine rev a little more freely. en I picked the demon- ator up it had only a couple undred miles on the clock l I kept the revs down to 00 in the gears and about 00 in top. As this corre- nds to a road speed of 70 h this arrangement was tty satisfactory anyway. e gear selector was not dded in and showing signs stiffness, making clunkless archanges almost impos- le. After the BM had its 0 mile service I gradually ed a little more throttle and rted to let the motor spin to 5000, which helped the archange quite a lot — or haps t doesn't make any 'erence to the selector, it ıld be that it's easier to ge the drop in revs and to chronise the change at her engine speeds.
As the motor worked its y up past the 1000-mile rk it became noticeably sier. Rolling the throttle to out half open when celerating would cause it spin up to about 6000, en earlier it had flattened at 5000. The gear shift came smoother too, even- lly producing clunks only en changing gear at ridicu- sly low engine speeds. At s point it was easier to nge gear without using clutch as this way the gs would mesh quickly and ently going up or down ough the box.
The clutch itself was ooth and took up the drive 'fectly, with the old BMW oit of raising the rear sus- nsion if the clutch was fed quickly. It did have a lot of vel between starting to e up the drive and being y home, which proved dis- ncerting when we tried me standing quarter runs.

Feeding the clutch in quickly gave good results but once it had started to bite it was tempting to assume it was home and drop it the rest of the way, which produced a most un-BMW-like wheelie. From our results it looks as if getting off the line was the most difficult part. The R90 ran consistently just over the 13 second mark with a relatively high terminal of nearly 100 mph — there was obviously enough go at the quick end of the strip and any improvements would have to be made on the start line.

While the riding position is superb, holding high speeds for any length of time restricts the machine to something in the 90s. I had thought that the little flyscreen would take more of the wind off the rider, which it seems to up to 70 or 80 mph, but above 90 mph you have to start hanging on. In fact after a few minutes at close to 100 mph, slowing down to 70 seems ridiculously slow, the riding position balances the wind pressure, the motor hardly seems to be working and there appears to be no noise at all.

The rocking light switch controls are very positive and easy to use, although if they were twisted round to be slightly more underneath the handlebar you would hardly have to move your thumb at all. The lights are good, but for motorway type use the dip beam cuts off too sharply not lighting enough road to be able to stop in from more than 55 to 60 mph. The clock proved to be a fascinating addition, especially on motorways where you can relieve the boredom by watching the signposts and working out when you should arrive at the next place. Listen closely and you'll probably hear passing BMW riders muttering "ten past nine, that must be Biggleswade" or words to that effect!

I spent most of the time on the BMW commuting from deepest Surrey up to the new MCM offices in the wastes of Peterborough, a journey which is at best boring, at worst painfully so. In fact the glorious month of February managed to produce freezing rain for every trip except the last and as the

The classic lines of the R90S engine. Basic design goes back 50 years but its smoothness is still hard to better.

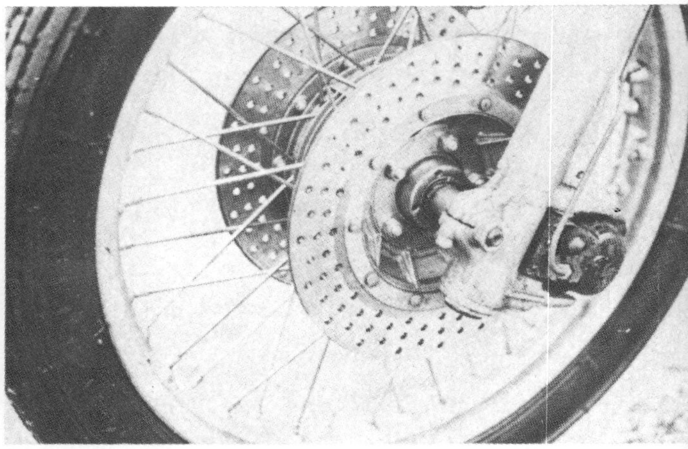

The twin disc front brakes are drilled to improve cooling and prevent water building up on the discs.

BM had to be returned I left early in the afternoon and actually spent two hours riding in a dilute sort of sunshine.

It was all very pleasant. I eased the R90 away from the offices, past a couple of cars who were trying to form a traffic jam and within a few minutes was rolling out on to the A1. All was peaceful, the only thing to watch was the odd east midlands driver — they all insist on getting into the right lane a good four miles before making a turn, unlike the more familiar metropolitan who never dreams of turning right unless he is in the left lane and vice versa.

But the BM can cope with a lot heavier traffic than that. Let it slip past a couple of cars, tuck in behind a truck, watch the road straighten out round the next curve and away, not even bothering to change down. Up to the Norman Cross roundabout, nothing coming, down two gears and heel over to the left in one long curve, the suspension mocking the cambered turns which go left, right and left again through the roundabout. Up to five and a half in the gears, past some lorries and the odd car . . goodness, is that five and a half in top? It didn't feel like it. What colour do they paint police cars in Huntingdonshire?

Left to its own devices the R90 would be quite happy at that, churning effortlessly along, shrugging off sidewinds, nonchalantly packing 95 miles into each hour. Easing out to pass something, a cat's-eye makes one of the wheels skip and I try to remember when I last

checked the tyre pressures. The clock is running twenty minutes slow, at first I put that down to a cunning trick by BMW to make me think I was getting places faster than I really was, but it's lost five minutes in the last week. Five minutes a week, that's twelve weeks to lose an hour, so it'll be right again in thirty-six months . . . the calculations are upset by a rapidly approaching roundabout, quick flip back through the memory banks to discover whether this is the one that's usually covered in rubber and diesel oil but slow down more than necessary anyway. The front brake buzzes as the pads try to play a tune on the holes in the disc, the nose of the BM goes down, the back twitches as I hit second gear a little too hard, steady it on the back brake and gently around the island where the surface turns out to be bone dry with enough grip to qualify for a drag strip.

Just as the R90 never lets you feel the rough edge of all the power it's putting down, so it is with the brakes. Even on smooth roads, heavy braking uses up all eight inches of front fork movement — you can see the marks on the staunchions, or at night it's almost equivalent to dipping the lights, but you never feel as if you're being tipped up on to the handlebars. The springs go down, the damping takes over and the R90 just stops, with perhaps a little squeal from the tyres.

The new disc brakes are certainly an improvement over the first type. They are better in the wet and although ours did manage a slight squeal occasionally there was nothing of the raucous noise which did unpleasant things to teeth, spines and old ladies and was symptomatic of the early BMW discs. On a long run, or in the rain, when the brakes are cold they take a little longer to come on, and after I'd hosed the BMW down it needed about a hundred yard ride with both brakes on to get normal braking back again.

But such is the nature of the R90S that you don't mind slowing down, its character takes all the urgency out of even high speed runs, you do things leisurely — and allow yourself plenty of time. Coming away from the roundabout the R90 accelerated briskly up behind a Triumph car. The driver was sitting quite contentedly in the fast lane with no other vehicle in sight. I could have flashed the lights, I suppose, but eventually he got round to looking in his mirror and pulled over. He probably wouldn't have noticed the lights anyway, after all there's only sixty or seventy watts up front. As for using the horn, I always reckon that when people show such a lack of other faculties there's no reason to suppose they're not deaf as well. But there are many occasions like this when one can and often does get annoyed, and when you get annoyed, ***you*** can do silly things too.

Runs down the A1 tended to alternate between a sort of elation, sweeping through the few fast curves, to a guilty conscience as I settled down on the straight bits and took time to read the instruments, followed by frustration as I tried to distinguish the blurred images in the mirror. This in turn was followed by a draught freezing down the back of my neck as I gave up and swung round to look behind.

Off the motorways, where the legal 70 mph is about as slow as the BMW wants to go, being restricted to 50 mph makes nonsense of the R90. It takes a lot of concentration to keep from creeping up way over the limit. Fuel consumption doesn't improve enough to show any real saving and the machine is generally wasted. The R90S is one of the most desirable and safe machines I have ridden but there's little point in owning one unless you intend to make full use of it.

BMW R90S

TEST CONDITIONS
Misty, damp patches on track. Ambient temperature 42 deg F. No wind.

PERFORMANCE

Maximum speed — 112 mph
Standing start quarter mile
13.6 sec; terminal 96 mph
Braking from 30 mph — 26 ft
Fuel consumption:
average — 45 mpg
hard riding — 43 mpg
gentle riding — 48 mpg
Minimum speeds in gears:
first — 3 mph
second — 6 mph
third — 9 mph
fourth — 14 mph
fifth — 16 mph
Speedo error; 2 mph fast at true 70
mph/1000 rpm in top: 16.2

HOW IT COMPARES

Model	Price inc. VAT	Max Speed	Average mpg	SS ¼ mile	Dry Weight lbs
BMW R90S	£ 1,899	112	45	13.6	446
KAWASAKI Z1 900	1,249	135	48	12.3	506
DUCATI 860 GT	1,299	110	42	13.2	452
HONDA 750-4	979	112	49	13.0	480
NORTON COMMANDO MK.IIA 830cc	928	110	50	12.5	410

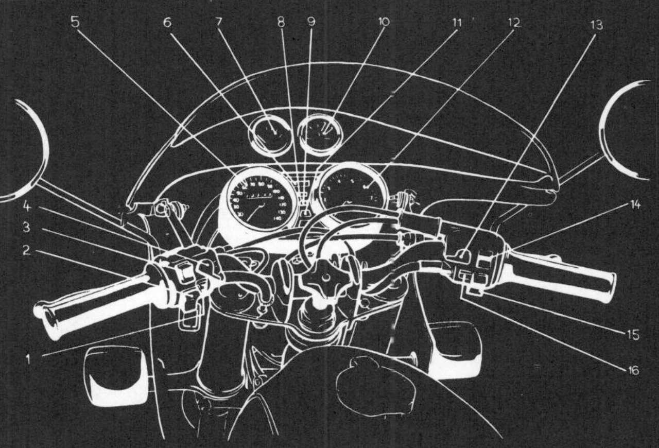

(1) ignition switch — side of headlmap; (2) main-dip-flasher; (3) horn; (4) lighting switch; (5) speedometer; (6) brake failure light; (7) voltmeter; (8) neutral indicator light; (9) generator/ignition warning light; (10) clock; (11) oil pressure warning light (turn tell-tale light below); (12) tacho; (13) ignition kill switch; (14) starter; (15) turn indicator; (16) twistgrip drum.

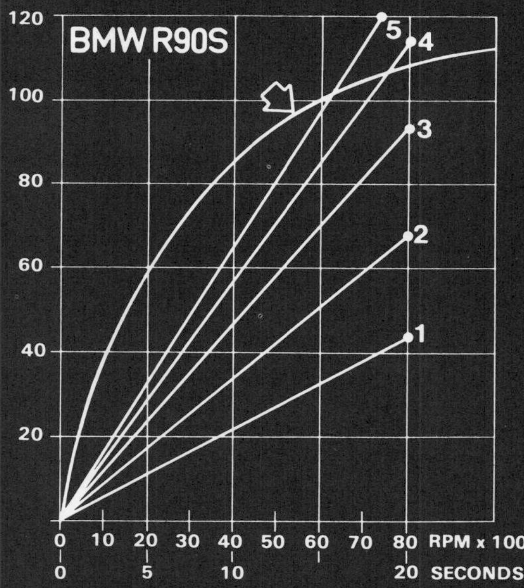

ENGINE

type	OHV, horizontally opposed twin
displacement	898 ccm
bore x stroke	90 x 70.6 mm
compression ratio	9.5:1
claimed output	67 hp DIN at 7000 rpm 75 hp SAE at 7200 rpm 55 lb ft at 5500 rpm
lubrication	wet sump, Eaton pump
carburettor	two Dell'Orto 38 mm
ignition system	twin contact breaker and coil
lighting/charging	238W Bosch alternator via solid state regulator to 12V, 25 a-h battery

TRANSMISSION

clutch	single plate, diaphragm spring
gear ratios	4.4; 2.86; 2.07; 1.67 and 1.50:1
primary drive	direct
secondary drive	Cardan shaft, palloid pattern crownwheel and pinion 3.0:1 reduction
alternative	2.91:1

CHASSIS

front tyre	3.25H19 Metzeler
rear tyre	4.00H18 Metzeler
front suspension	two-way damped telescopic fork full movement 8.2 inches
rear suspension	swinging arm, 3-position dampers full movement 4.9 inch
castor	not available
trail	3.5 inches
wheelbase	58 inches, laden
front brake	twin 260 mm disc, hydraulic operation
rear brake	200mm sls drum
overall width	29.1 in
overall height	47.6 inches
seat height	32.3 inches
overall length	85.8 inches
ground clearance	6.5 inches, laden
dry weight	446 lb
test weight	474 lb
tank capacity	5.3 gallon, including .44 gallon reserve
oil tank capacity	4 pints

PARTS PRICES (inc. VAT)

front mudguard £50.13
handlebars £11.04
speedo cable £3.89
exhaust pipe (each) £27.83
set of points £3.83
set of pistons (complete) £25.49
list price .... £1899+ £10 delivery
warranty
Concessionaire: BMW UK Ltd, 994 Great West Road, Brentford, Middlesex.

‘Astounding performance for a 750....’

BY DAVE MINTON

I CANNOT but help think that BMW's policy of concentrating promotional and marketing efforts around the "non-motorcyclist" has cost them dearly. For however admirable the principles behind the scheme, and however beneficial to motorcycling in general the rewards of spreading the good word might have been, I have a gnawing suspicion — entirely without factual identity incidentally — that a good many motorcyclists have turned away from the prestigous twins because of the overpowering charisma of opulence and luxury surrounding their contrived image.

That they are better built than the majority of their competitors is not in doubt. A quick check through the sound of the running engine, control operation, casting quality, enamel and chrome finishes, is pretty conclusive evidence of BMW's high aims at real quality. You only have to sit on one to realise that the bike range is one of the few ergonomically designed available, and for me that counts for more than all the cylinders and bulging pipes in the world.

In other words, BMW, definitely subscribe to most enthusiasts' ideals of the most practical form of motorcycle owning there could be. Time was when their price put them away out of reach of most riders, but now, at £1,400 for the R60/6, the subject of this appraisal, any number of alternative big bikes, all selling like hot cakes, exceed it by tens of pounds a go.

Instead of getting into the newest lump of chrome around we thought it might prove interesting to size up a well established part of the scene in a new light. Because of the rich young exec's fun machine image BMW have quite deliberately chosen for themselves, we wondered how the smallest BM would stand up to scrutiny as a bike buff's sporting/utility all rounder. The answer is short and simple — very well indeed. It's an extremely practical motorcycle.

Obviously, for less money any number of alternative 750s abound, but as owners of these bikes will know to their cost, that unless they have shaft drive, in which case their price is very high, or full rear chain enclosure — very, very rare indeed, if they are driven at the speeds for which the machine was intended and the price of a new chain every 3,000 miles or so (around £10 a go) works out very expensive. At that rate, if you cover only 10,000 miles annually, you are faced with a £33 bill come birthday time again. And by the time you've bought the spare links and chain repair kit and a spot of special lubricant you could reckon on £40. Spread that over three years with the addition of a new set of sprockets plus numerous other transmission requirements connected to chain maintenance and you'll be lucky to get off with a total of £140.

I cover between 30,000 and 40,000 miles each year, and have done for the past 15 years. Keeping to my calculations would result in a cost of 1,750 pounds on chain alone! It matters little how you look at it or how many machines are involved, one or a dozen, it comes out exactly the same each time. Shaft drive has distinct advantages considered along those lines.

Anyway, 750 or 600, it's unlikely that many riders are going to need more speed than the R60 puts out.

One of the disadvantages of riding little but press machines is the difficulty in obtaining fully run-in motorcycles, consequently the kind of performance that long term private owners can anticipate from their own bike often escapes the press. Also because of the familiarity of handling a lot of exotica we, and I speak I am sure for my fellow testers as well, have to make a concious effort not to fall into the trap of becoming blase about anything not absolutely staggering. Put those two together and maybe you will accept the truth of my careless reception of the R60. "A 600", I thought. "So what's in a 600 these days?"

Firstly I saw by the big mileage knocked up on the R60 that it was a well used one. Slack it was with that clickety whirring common to all well-used pushrod engines. Not noisy mind, but loose jointed and so obviously familiar with speed, as the blue and brown sections of exhaust pipe close to the barrels showed. As a matter of interest I was ticked off by a young Honda owner for "riding the poor thing too hard" as he noticed the discolouration. It's a falacy, and one that is unfortunately gaining ground because of Honda. All air cooled engines will discolour their pipes if ridden as the makers intended them to be. The only ones that will not are Hondas because Honda avoid the tell tale mark by using two pipes, one within the other. The inner one takes the brunt of the flame's heat and leaves the outer one sparkling clean.

The well worn in engine was new to me. Never previously have I had the chance to wrap a fist around anything but new BMWs, but this one responded to the throttle with an eagerness that surprised me, running over the top rev limits so easily I had to watch the rev counter carefully, and within an hour of collecting the machine I had forgotten all my good resolves about sensible riding to obtain maximum economy, and forgot too my disdain of "little" engines.

The smoothness of the engine exceeded that of its bigger brothers considerably. Whereas the R75, for instance, collywobbles to around 2,500/3,000 rpm very positively, this one's movement was lighter, fading out sooner, which suggested a more flexible machine. BMW's generally overemphasised lateral rock, caused by the secondary rocking effect of the otherwise naturally well balanced opposed piston layout, hints often at inflexibility. I found the engine rock gentle enough to overlook at any but imprudently wide throttle openings at irrationally low revs. Used intelligently, the engine provided ample torque in top gear from around 1,800rpm (28mph) to return a relaxing ride. Lower engine speeds, right down to 1,200 revs (19mph) in top still provided a useful performance but the discipline involved required more concentration than the use of a lower gear ratio.

A side effect of wafting along on such diminutive throttle opening in top gear was to frighten careless pedestrians out of their wits who quite apparently assumed the lack of noise meant a lack of approaching vehicle. Four wheelers are accompanied by so much more noise

from the broad and loaded tyres involved, so the silent bike did little to improve mototrcyclist/pedestrian relationships as those involved generally deplored the silence as dangerous!

I generally changed up through the gears at around 3,000rpm if I was not hurrying, and through city traffic kept pretty much to this in the intermediates simply to make the best use of the bike's super traffic agility. At this engine speed, acceleration was snappy enough to keep ahead of developing "situations" long before they arose. Road speeds of 17, 24, 34, 42, and 47mph resulted, and off the motorways and dual carriageways; providing I was willing to hold speed down to the legal limit and so provide every other roaduser with an unnecessary hazard as he overtook, the fuel consumption of the R60 dropped to an incredible 90mpg. Useful, commendable even, one might say, predictable with a correctly engineered four stroke of this type, but what a crashing bore!

More real speeds, incorporating the illegal, but normal traffic pace, along dual carriageways and motorway going provided a much more realistic petrol consumption. Using what I term progressive acceleration — being the type whereby the twistgrip is never allowed to advance beyond engine speed requirements any more than is absolutely necessary for brisk, but not hard acceleration — up to 5,000 revs, including cruising speeds at that in top gear (close on 80mph) provided a highly commendable 66mpg. Even harder riding never really caused any great concern of fuel use, the worst I achieved being a supremely exhilerating 49mpg cross country run, during which time I changed up through the gears on instinct rather than any instrumental orders. 6,500rpm seemed about the optimum. More did encourage a mite more power but I prefered not to abuse what was an altogether smashing engine.

As is BMW's mormal style, acceleration was considerably quicker than the information radiated by the bike due principally to the rock sure method of delivering it.

Far and away beyond mere acceleration ability was the R60's appetite for high average speeds. Why some motorcycles have it and some don't is one of those personality traits that provides lie swapping bikers with as much meat as Hailwood ever did. But whatever, the "little" Bee Emm had it right down the line.

As much as anything else I feel sure that the riding position had much to do with it, coupled with the sheer simplicity of riding fast. A lack of diturbing vibration, a wide power spread so reducing gear changing to a minimum, light clutch, supremely comfortable suspension, and vastly improved stability — about which we discuss shortly.

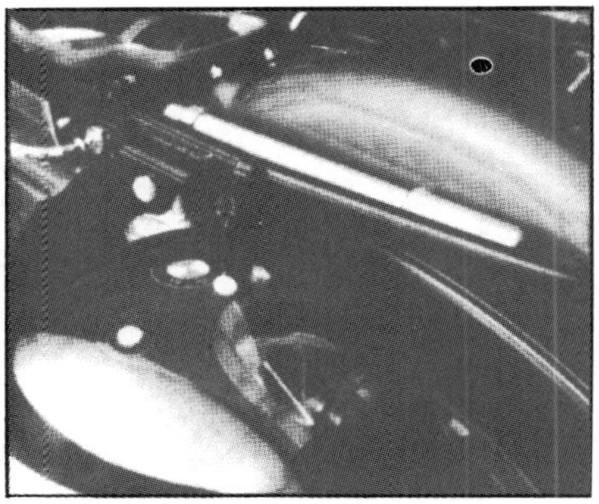

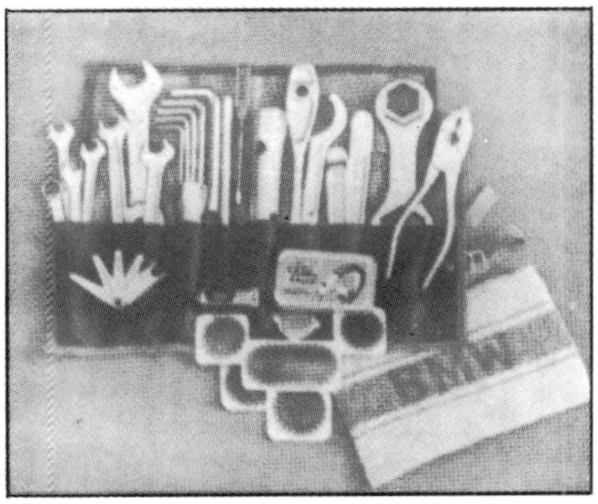

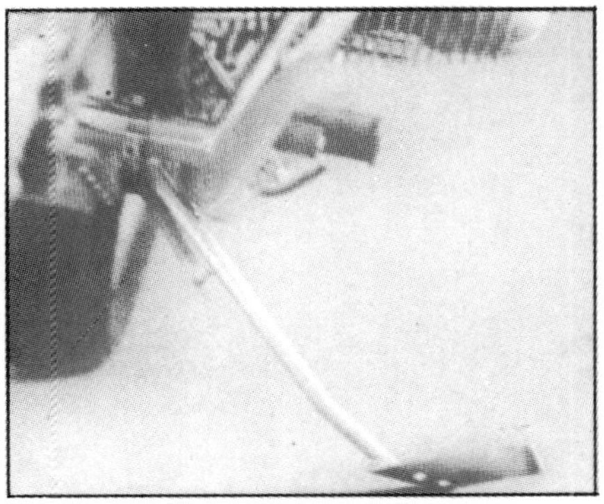

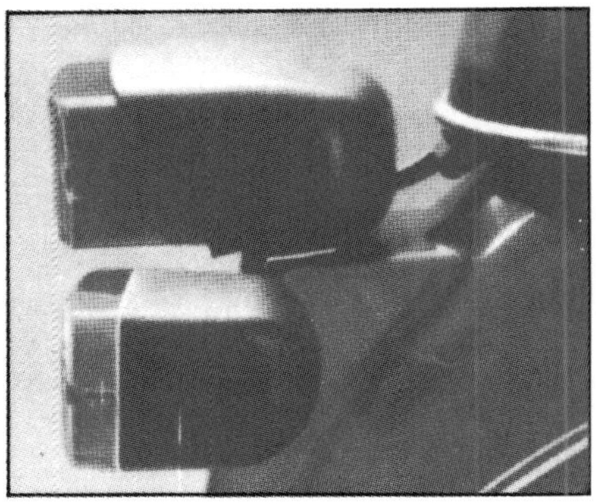

The point is that the makers suggest 6,400rpm to be the optimum engine speed, at which period in top gear the bike is loping along at 100mph. The rev counter and speedometer more-or-less confirmed this, but whatever cruising speed I required, right up to top speed suited the 600 exactly. By this I do not mean that like, for instance, a Trident, or an H2 Kawasaki, or an MV Agusta four, it will hold its top speed for a long time and therefore prove ultimately faster than the 600 BMW. Faster those three might be, as examples of many others like them, but the genuine top speed cruisers are rare birds. Bikes that will pound along tirelessly at full chat for a whole day and finish fly spattered, blue piped, and idling gently without a ping or a tick to the heat warped engine as you switch off.

I was impressed. For a 600cc engine this non-sportster, usually regarded as more of a "lesser" '75 or '90 than a bike in its own right, was undoubtedly faster over a big mileage than a good many of the "faster" and more powerful 750s around. Yet it managed it without losing a fraction of comfort, tractability, silence and sheer undemanding riding simplicity.

This in turn leads into the stability factor. R75/5s have given me grey hairs on occasions because of their unpredictable habits at speed. R75/6s are much better, but something has puzzled me about them. Why did the R90/S I rode behave so much better than the R90/6, although I actually preferred the /6 model, stability apart?

Both had the long wheelbase and were apparently identical, yet one was positively a better machine than the other. If you read through the last BMW test we carried (see MCI January 1975) you will see some of my comments regarding its high speed instability. It's a fact that BMW have been carrying out some intensive research and development into high speed stability, hence the various steering geometry changes, and wheelbase lengthening exercises. The latest improvement has been to strengthen the spine of the machine by force-fitting an internal tube along the inside length of the present one. Large diameter wheel spindles are also part of the scheme to improve rigidity further still. I have no doubt that the lighter turning mass of the 600 engine, compared to the 900cc one, assisted the lessening of frame stress, but for all that the improvement I felt in the R60 proved conclusively that BMW are making very real improvements to their machines' safety at speed. The way I was able to fling that bike around in complete safety was nobody's business but mine. The only interruption of bend-swinging rhythm and flow was when negative power was under way — neither roll on nor off but just enough to see the bend through safely — you know the situation I'm sure. Then with everything heeled well over, something, such as a feathering of throttle of brake or a dip in the tarmac would set the back end alone yip-yip-yipping. Not much and it never deteriorated and was always disciplined by the rest of the bike, but I still feel that until those unbraced pivoted fork bearing points are properly braced things will never improve. That silly bolted on sub-frame does nothing at all.

That it was good was echoed by a bike dealer, who, innocent of my visit, commented about the "state of the bleedin' sidewalls on that flamin' BM's tires!" It might have looked crazy, Bruce, but it was *that* good. Believe me. As far as cylinder head scraping is concerned, forget it. I've said it before and I'll say it again because riders who should know better insist that clanging down the rocker boxes is all part of the fun, but it simply aint on. Once or twice at low speeds I have touched gently during a gentle direction flick change, but at speed I always find that because of suspension crushing, my feet and the footrests touch down first. Start putting the bike down far enough to touch the engine and trouble will be your way in less time than it takes to say it.

One of my greatest delights was to get back to a good drum brake. The experience was so good it hurt. Ah! The sweet bliss of a gentle anchor, biting deep but so delicately and with such feeling, into speed. No dead hydraulic pressure, void of sensitivity and helplessly impotent at the first douche of water or road dirt. a fine old fashioned, weatherproof drum brake with a balanced shoe control. Not as powerful as twin discs; maybe not as powerful ultimately as one disc, but so very, very predictable and trustworthy. Think you, of the last time you actually *needed* the immense power of your massive discs during an emergency halt. For me at least the possibility is as frightening as the result of such a happening. I doubt I could control a bike with crushed and bending front suspension under the terrific strain of braking like that. And could I actually feel the point just below wheel-locking in such a panic grab? i think not. I prefer to stake my life on brakes such as these. This one did all I wanted of it, plus a little extra in reserve. That's enough for me.

I liked the finish of the black R60. Thick, durable glossy paintwork, smooth, liquid finished chromework, with not a buffing graze in sight wherever used, and that sparingly. Top grade cast aluminium clean of laquer and varnish so you know how good it is. And — laugh if you like but think before you do — strong, well made plastic components. I'm a strong supporter of plastic on bikes because the material neither weathers, oxydises, rusts, peels or accepts permanent shallow blemishes. What's more it's less expensive that its metal counterpart would be, and probably no less strong either.

Yeah, I know the front mudguard costs over fifty quid, but that's just the single silly price that everyone ridicules. Would you really not buy a BM because of that? I'm pretty sure I could find a way round it if the problem was mine. It isn't, but I wish it was.

As stated previously, the switchgear was unbelievably thoughtlessly designed compared to last year's traditional perfection. The headlamp was just about all any rider could wish for, being a big and powerful quartz halogen powered broad white beam of such power you could see erring car drivers cower under its gaze.

To return to the openeing theme. is it a bike for the active, dedicated bike buff

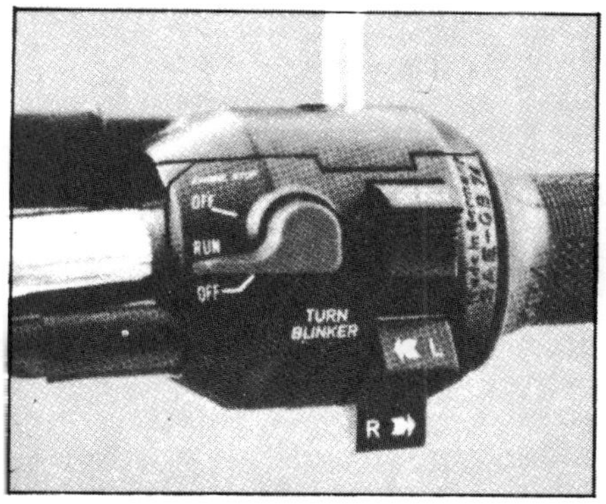

who needs transport, who loves good riding but who has for all that a budget to live by? I think so, especially for the guy, or girl, who buys for keeps. Over a period of five years plus I can think of nothing more practical.

Smooth, economical, reliable and durable. What more could you want?

After a run back home along one of my favourite roads through the empty countryside of Worcester and Herefordshire I caught up with and passed a guy on a 750 Honda. He latched on behind and we unravelled the deep evening miles together on our murmuring bikes. I noticed in the BM's twin mirrors that he dropped back on the twisty bits, catching up slightly on the straight runs, but not enough for us to arrive simultaneously at the next town on. I pulled into a garage to tank up for the next day's riding and he followed.

That's when he came out with his congratulatory "My word that was good. Astounding performance, your 750." I broke the news about my "750" as gently as I could to him. I saw the gleam in his eye change to a glint as the message sunk home.

SPECIFICATION BMW R60s6

ENGINE

Type: Horizontally opposed air cooled twin. All alloy with iron barrels. Car type construction with all plain bearings.
Valve operation: Pushrod from single camshaft beneath crankshaft.
Capacity: 599cc.
Bore and stroke: 73.5 x 70.6 mm.
Compression: 9.2:1.
Carburation: 2 x 26 mm mechanically operated Bing instruments breathing through paper filter.
Electrics: 12v 25 a/h battery charged by front crankshaft mounted 280w alternator. Coil and contact breaker ignition.
Claimed power: 40 bhp (DIN) at 6 400 rpm. 35.4 ft/lbs torque at 5,000 rpm.

TRANSMISSION

Clutch: Dry single plate diaphragm spring type.
Gear ratios: 1st 14.78:1; 9.60; 6.95; 5.61; top 5.04:1. Selection by left side one down and four up foot lever.
Final drive: Shaft via tunnel in right side pivoted fork leg and incorporating rubber vane shock absorber on gearbox end.

FRAME:

All welded steel tube full loop duplex cradle. Single large diameter spine tube reinforced along centre with second tube. Bolted on rear sub frame. Taper roller steering head and pivoted fork bearings.

SUSPENSION

Front: BMW telefork with two way damping and 7.87 in travel.
Rear: Pivoted fork with two way damped, load adjustable units and 4.9 in travel.

WHEELS

Front: 3.25 in Metzeler high speed ribbed tyre on WM3 alloy roadster rim. 7.87 in (200 mm) 21s drum brake with self centring shoe operation.
Rear: 4.00 in Metzeler studded high speed tyre on WM3 alloy roadster rim. 7.87 in sls drum brake.

INSTRUMENTATION

Matching illuminated speedometer and rev counter. Warning lights include oil pressure, charging circuit, turn signal, neutral, brake servicing, and main beam.

EQUIPMENT

Centre and prop stands. 3.96 gallon steel fuel tank including .5 gallon reserve. 4.84 gallon tank optional. 7 in 60w quartz halogen headlamp. Twin helmet locks under lockable hinged dual seat also containing tool kit and document compartment. Three position hydraulic steering damper. Passenger grab rail. Twin mirrors.

DIMENSIONS

Ground clearance: 6.25 in.
Wheelbase: 57.7 in.
Seat height: 31.9 in.
Kerb weight (as tested) 461 lbs.

PERFORMANCE DATA

Fuel consumption: Mild 50/70 mph cruising, 69.5 mpg. Fast riding, 50 mpg. Average around town 68 mpg.
Fuel quality required: 4 star used throughout test period.
Oil consumption: 1444 mpp.
Highest speed attained: 102 mph with 217 lb rider wearing touring leathers. Best practical upright speed, 93 mph.
Gear speeds: at 6,500rpm — 28:50:71:88:99mph
Speedometer accuracy: Accurate at 30 mph. 1 mph fast at 60 mph. 3 mph fast at 90 mph.

Warranty period: 6 months or 6000 miles.

PRICES

Retail price new machine:	£1,399
Piston and ring set	5.51
Valve spring set	96
Clutch fibre plate	10.87
Speedometer head	28.03
Contact breaker points	2.77
Fork staunchion	30.73
Exhaust pipe	20.14
Silencer	29.81
Throttle cable	1.47
Light unit	9.32
Fuel tank	76.90
Brake linings	4.35
Cylinder head gasket	47
Handlebar lever	4.15

BMW's R90S
THE WORLD'S BEST BIKE?

THERE IS universal agreement amongst motorcycle enthusiasts that the BMW is about the world's finest motorcycle. The marque still seems to set a standard all but unapproachable by most makes and totally out of sight to others. And if the blue-and-white BMW nameplate guarantees something above and beyond the norm, then the top model in the three-bike range ought to be the greatest thing built with a wheel on each end. TWO WHEELS looked at the latest R90S sports tourer — almost a Mk II version because of several detail changes — for justification of the legend.

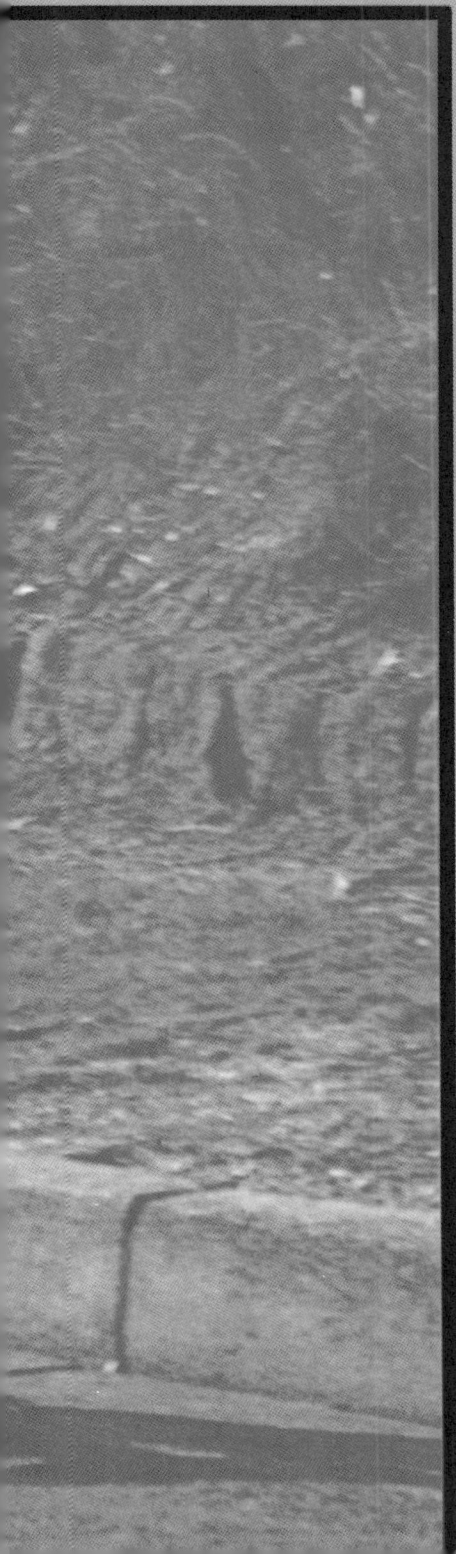

There's no mistaking the capacity — the sidecover script sees to that! The company reportedly expect at least half their sales to be in the 900 class.

THE R90S combines beauty of design, stunning overall appearance, excellent — and mostly exclusive — appointments, complete instrumentation and reasonable outright performance. It is not the fastest on the road, nor does it accelerate anywhere near as dramatically as some. For that matter its engine is by no means as sophisticated as many and its exhaust note is hardly exciting . . .

But it combines so many virtues which are its own (and some other makers') that it comes out in direct comparison with its contemporaries as something very special indeed.

Of course, at a price perilously close to four grand on the road and fully equipped, the thing ought to be out of sight! And it should stay there forever, because it would probably be the last machine a rider would buy after running the full gambit of a whole range of "ordinary" road irons.

The marque has always been

Tank removal reveals brake fluid reservoir and brake light sensors — plus a backbone of impressive dimensions. Instruments succeed in appearing to be one-piece panel, but that's not the case. The dials and warning lights clamp atop the bars in conventional fashion. The clock and voltmeter are integral with fairing. Panels either side of fairing dials contain handy service information.

BMW's R90S THE WORLD'S BEST BIKE?

The powerplant idles with some lumpiness, but becomes immediately smooth with throttle. The bike has an extremely quiet exhaust note and header pipes are joined at front of crankcase. End plates remove for access to points, generator and electrics.

exclusive, if only because of its price, and its sales have tended to be to the older enthusiast more able to afford it. As a result, BMW has not enjoyed much of a reputation for building road-burners — until recently that is, when the range was updated in the most radical way — by a total change in design.

First of the all-new BMW range (then, as now, confined to just three variations of engine capacity on a common basic design) was introduced in 1969. The horizontally-opposed twin with shaft final drive concept remained, of course, for BMW has been committed to this design for 50 years. But the new unit was so radically different that only the rockerbox caps were common to both designs and were totally interchangeable!

The range consisted of the previous capacity models of 500 and 600 cm^3 but included Bee-Emm's first recognised superbike, the 750 cm^3 tourer. Halfway through 1973 the German factory made a seemingly minor modification to the range when it endowed the marque with even better handling by increasing the wheelbase by just 5 cm. It achieved this feat by simply lengthening the machine's swing arm by that amount and fitting a suitably longer Cardan drive-shaft.

At the same time, the factory also angled the carburettors more towards the engine castings to allow more room for taller riders' shinbones, and cunningly concealed this by saying the extra legroom was made possible by the longer wheelbase. The ploy probably worked. At least one road-tester remarked that it was about time the Germans allowed more room between leg and inlet manifold, and applauded the longer wheelbase frame for achieving this end! He also remarked that the factory claimed handling was unaffected by this important detail modification, and he agreed there was no discernible difference between pre- and post-long wheelbase models.

How wrong he was! The modification was done in readiness for the belated arrival of the sports model — the all-new R90S.

The modification improved handling dramatically, not perhaps at the fairly reasonable speeds one would expect the "normal" BMW to be ridden, but at the top end of the machine's performance where many of the owners (though perhaps not in Australia) might reasonably be expected to be riding. There were many rumors circulating at the time about a secret new model the Germans were soon to introduce to the Bee-Emm range, with speculation running rife about flat-fours, special overhead camshaft heads, 1000 cm^3 engine capacities and more, much more.

When the news broke it was exciting enough, although for many it was something of an anti-climax because the new models continued to use the old-hat overhead valve engine; there wasn't much of interest in that! But there were radical changes for the Bavarian marque, because the factory had seen fit to follow fashion (which you could never accuse them of doing before) by fitting five-speed internals to their gearbox and introducing two 900 cm^3 variants.

In doing so, the factory was forced to drop the smallest capacity model, the 500 cm^3 R50. Its main disadvantage lay in the fact that it shared everything with its bigger brothers except engine capacity and performance. The device was only some 30 kg lighter than the bigger models! With the accent swinging very much towards superbikes and with BMW's thinking enjoying the benefit of a new, young philosophy, there was no place in the range for the "small" 500.

The new drilled front discs. We're still not sure that the holes do anything other than reduce unsprung weight!

And so the updated (you could hardly call it new) model was introduced at last, ending months of speculation as to just what exciting new things BMW had come up with for their new flagship.

Not everyone liked it; staunch "Boomer" fans decried the semi-cafe racer look of the sleek, low handlebar fairing and there were some who threw up their hands in horror at the indecently naked fork legs. Still others remarked on the gaudiness of the Christmas-tree dash panel.

But history tells the story very graphically, for such was the unprecedented demand for the new Bee-Emm flagship that we had to wait almost six months before the first of a thin trickle of R90S models reached Australia — to be snapped up eagerly by riders whose names had burned holes in dealers' order books for months.

Cosmetically, the new sports mount was in there pitchin' with the best of them, and the new paint job of black relieved by air-brushed silver smoke set off the promise of engine specifications in a most exciting way.

The later model has the same basic specifications as the first model, but is now available in a very smart so-called Daytona Orange (milky gold *might* be closer) as well as the original color scheme; it also offers a totally new — and initially confusing — space-age switchgear and perforated brake discs. It is at once a very striking machine and one cast in the mould of an out-and-out sports model, its low, racy lines accentuated by perfect paintwork and gleaming chrome.

The frame is, of course, trad black baked enamel, unique in using oval-section tubular steel. It uses full-cradle double-loop construction with closed ends and argon-arc welded bracing at major stress points around the steering head and swinging arm pivot points. The main tube is reinforced with a close-fitting round tube enclosed within its length. The box section frame is extremely rigid in all planes, but, sadly, has the rear sub-frame assembly bolted on just under the nose of the seat and above the pivot point for the swing arm.

Lockdown seat has strong hinge structure and there are two bins for carrying gear. One fits over the battery (removed in this shot) and another is fitted under end of seat.

BMW's R90S

The sub-frame forms the base for the dual seat and the top mounting point for the rear shock absorbers; it is hung cantilever-like from the two mounting points which, in themselves, are well triangulated, but there is no way in which a bolted-on rear assembly should be used on a motorcycle. It works perfectly well, but there is at least one instance known to this publication in which the bolts have worked loose in service. It was during a two-hour production race at the Calder Raceway in Melbourne, when Bob Rosenthal's R90S was side-lined for several laps while replacements were found for a couple which had disappeared during the race!

Why, then, is the factory staying with this design?

We suspect it is to help defray the cost of replacing a damaged frame assembly — there can be no other justification. To be fair, there has been little trouble with bolted frames and BMWs are not the only examples of this design quirk. Perhaps in normal servicing a cautious owner should just give 'em the occasional tweak to check, avoiding possible trouble?

Long-travel front forks boast 200 mm total movement and are of the fashionable exposed-tube pattern, with the lower sliders sealed by longer rubber sleeves than most. The sleeves include felt seals soaked in oil for automatic cleaning of the fork tubes. Two-way damping is (naturally) featured, BMW developing its own special damper fluid. The forks are fitted to the frame through a huge alloy bridge-piece, the centre stem pivoting on tapered roller bearings.

Hydraulically-operated twin discs are fitted to the front, the discs perforated to combat brake fade in the dry and moisture building in the wet. The caliper mounting bracket is cast as one piece with the lower fork slider and adds the bonus of extra rigidity to the fork itself. Cleverly, the brake fluid master cyclinder is mounted on top of the crankcase under the fuel tank; nicely out of harm's way, but it also means the tank has to be removed to inspect the cylinder.

Directly fitted to the handlebars is a prominent handlebar fairing which BMW refers to more accurately as a "cockpit" fairing. It dominates the front end of the machine, adding greatly to the overall sports image and dramatically reducing wind pressure.

But the appointments don't end there, for the fairing also contains two instruments — a clock and matching voltmeter — fitted as standard equipment for the first time ever on a motorcycle. Also included are two very handy specification panels which list most of the essential data on tyre pressures, oil capacities, tappet settings, ignition timing and other information generally contained within the handbook you can't find because Fred has borrowed it.

A dash panel is clamped to the top of the handlebars and sits in position below the lip of the fairing. The panel contains matching speedo and tacho, with a row of lights between them to indicate brake fluid level, neutral, battery charge, oil pressure and blinker. The main beam warning light is mounted in the rev counter — which is red-lined at 7200 rpm — and the speedo includes a trip meter. It has got to be the best dashboard setup around and that says a lot in view of some of the excellent Japanese designs.

The new switch gear almost warrants a report on its own. Initially confusing — which is odd because they are very well labelled — they are totally different from any on Japanese machines, including not only their operation but their positioning.

The right switch includes the starter button, a rocker blinker switch and a cut-out switch, while the left includes main light switch, horn button and a rocker switch for dipping the main beam. The handlebar clamp-cum-switch body is finished in matt black with clear white lettering, the cut-out in a contrasting red and the main light switch in bright yellow. Very neat and easy to assimilate. Thankfully, the Germans have seen fit to run the wires neatly alongside the handlebars, ignoring the absurd Oriental — and sometimes Italian — practice of enclosing the wires within the bars themselves.

A sensibly-sized 24-litre fuel tank blends well with the long, contoured dual seat which swings up to uncover the large 12-volt battery. The battery sits under a moulded toolbox containing the world's finest motorcycle toolkit, *and* your very own face-flannel with BMW woven into it! How pretentious, and yet how typical of a machine which is designed with so many little features built into it that it is damn near impossible to list them all: Like the centrestand, which is so well positioned it holds the machine with either wheel clear of the ground. And the wheels are designed so you can remove them by merely loosening fork clamps. We found the wheels can each be removed in two minutes — and replaced without the need to line them up again. And there's a tyre pump, which hides demurely away under the seat until needed, the prop stand which flicks up when the bike is lifted upright, and the three-position damper knob which locks the hydraulic steering damper in city, touring or road-burning positions!

The engine design is also such that routine maintenance is very simple. With the pots out in the breeze, good thermal efficiency is guaranteed, with tappet adjustments needing only the removal of

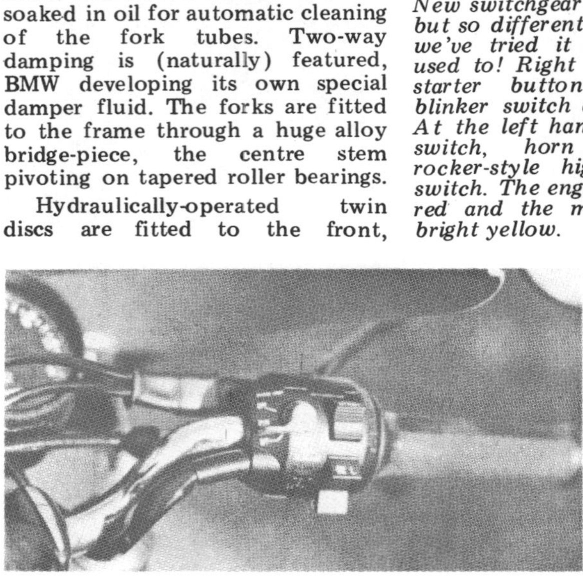

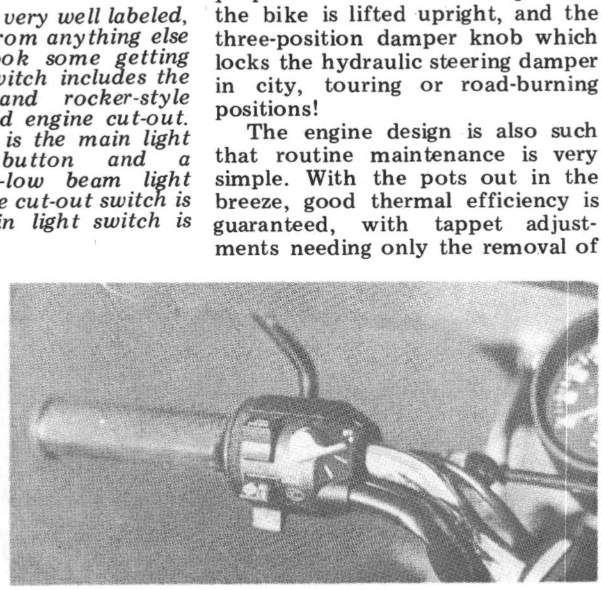

New switchgear is very well labeled, but so different from anything else we've tried it took some getting used to! Right switch includes the starter button and rocker-style blinker switch and engine cut-out. At the left hand is the main light switch, horn button and a rocker-style high-low beam light switch. The engine cut-out switch is red and the main light switch is bright yellow.

a centre nut and two small "nuisance" anti-distortion studs to lift a rocker cover and expose the tappet screws. A huge casting on the front of the crankcase can be removed by unscrewing three captive Allan screws, uncovering the contact points, auto-advance unit and 280 Watt three-phase alternator mounted just inside.

A rubber bung on the side of the crankcase casting just above the left carburettor is lifted out when ignition timing has to be checked, and static setting can be determined by the time-honored cigarette-paper-in-the-points-till-it-is-released-as-they-open trick, or by strobe light if you must be like that. Similarly, the two Dell'Orto 38 mm carburettors hang out of the frame where they can be readily adjusted. In keeping with the sporting performance of the Bavarian twin, the carbs include large capacity float chambers and accelerator pumps.

Suitably long life for major engine components is ensured by adopting automotive practice in the design of the chunky one-piece forged crankshaft with massive bearing areas and plain bearings fed by a high-volume oil pump. A disposable oil filter element is fitted in the system, just above the wet-sump oil container which is a large neatly-finned pan bolted under the centre of the crankcases.

We expected the BMW not to be quick off the mark, but it certainly has all the useful acceleration a rider needs. Cornering performance is extremely impressive!

Bearing shells are the same size as the split plain bearings fitted to the BMW six-cylinder cars, so there should be little trouble with them. If they need to be replaced, the barrels are lifted merely by removing four nuts on the long through-bolts. The connecting rods can be dismantled without removing the engine from the frame! Likewise, the gearbox can be removed for servicing without disturbing the engine too much — although it's a fairly major undertaking. Thankfully, the later BMW models have overcome the spate of gearbox problems which plagued the earlier five-speed boxes just after they were introduced.

It has been pointed out in detail, by many people who have probably never ridden one of the models,

SPECIFICATIONS

MAKE ... **BMW**
MODEL ... **R90S**
PRICE ... **$3495**
ENGINE: Air-cooled ohv four-stroke twin. All-alloy, cylinders horizontally opposed, mounted transversely in frame. Forged one-piece cranksaft, plain big-end and main bearings.
Bore x stroke ... 90 x 70.6 mm x 2
Capacity ... 898 cm^3
Compression ratio ... 9.5:1
Claimed Power ... 50 kW (67 DIN hp) at 7000 rpm
Lubrication: Wet sump for engine, Eaton oil pump, replaceable oil filter cartridge. Separate oil supply for gearbox, final drive shaft, crown wheel and pinion.
Carburetion 2 x 38 mm Dell'Orto PHM. Paper element
Ignition ... Battery and points, coil
Starting ... Electric, no kick starter fitted
TRANSMISSION: Diaphragm clutch, crankshaft-mounted, to five-speed constant mesh gearbox. Left side gearchange, one down, four up. Final drive by enclosed Cardan-shaft to sealed crown-wheel and pinion.
Overall ratios (with standard 3:1 final drive):
First ... 14.2:1
Second ... 8.58:1
Third ... 6.21:1
Fourth ... 5.01:1
Fifth ... 4.5:1
Alternate final ratio available ... 2.91:1
FRAME: Welded box-section oval tubular steel, strengthened and stress relieved. Rear sub-frame bolted into position.
SUSPENSION:
Front: BMW telescopic, with coil springs and two-way hydraulic damping. 200 mm travel.
Rear: BMW spring/shock units, three-way adjustable mounted on swing arm/drive shaft housing. Swing arm pivots on adjustable tapered roller bearings.

WHEELS, TYRES AND BRAKES:
Front: Lipless alloy rim, 325 x 19 Metzeler ribbed tyre, perforated steel twin discs with floating calipers, 260 mm diameter.
Rear: Lipless alloy rim, 400 x 18 Metzeler high-speed block. Drum brake, single leading shoe, 200 mm dia.
ELECTRICS: Crankshaft mounted three-phase alternator, 2800 Watt. 12V/25 Am. hour battery. 60/55W. QH headlamp, 10/25W stop/tail.
DIMENSIONS:
All-up weight ... 215 kg (474 lb)
Wheelbase ... 1465 mm (57.5 in.)
Overall length ... 2180 mm (85.6 in.)
Overall width ... 740 mm (29 in.)
Overall height ... 1210 mm (47.5 in.)
Seat height ... 820 mm (32.2 in.)

PERFORMANCE:

ACCELERATION:
0-100 km/h ... 4.9 sec
Standing 400 m ... 13.2 sec
Maximum speed ... 202 km/h
Fuel consumption (overall) ... 17 km/l (51 mpg)
BRAKING:
From 50 km/h ... 9 m
From 100 km/h ... 40.8 m

HOW DO THE FINER POINTS SCORE
(Out of 10)

Instruments ... 10
Comfort ... 8
Tool kit ... 10
Finish ... 9
Vibration ... 9
Lighting ... 9
Positioning of controls ... 9
TEST BIKE SUPPLIED BY: Ron Angel Holdings, 635 Bridge Road, Richmond, Vic.

BMW's R90S

that the shaft-drive flat twin suffers from an unsettling torque reaction when braking, changing gear or backing off suddenly when cornering hard. Sounds logical enough, particularly as you can produce the reaction by tweaking the throttle while the machine is stationary; the bike *does* tend to flop over in such a situation. But that changes when the thing is mobile.

In fact, the only limiting factor on the machine under test was the crashbar, which wore itself down quite happily when we indulged in some pretty lurid cornering. It should be stressed this could only be done deliberately and on corners which we knew our way around pretty well. On the open road at any sort of speed there was no way they could be made to ground — and it seemed pointless to try!

The Dell'Orto cold-starting system uses a choke lever in the normal Bee-Emm position on the left of the crankcase near the nose of the seat, but differs in working more like an accelerator pump squirting the fuel in. It was necessary to close the control partly the moment the engine fired up, or uneven running would start almost immediately.

It idles with a degree of lumpiness, but becomes almost uncannily smooth the moment the throttle is opened and load applied. It remained smooth, except for a small vibration period at just on 4000 rpm, right through the range.

Another old characteristic of the BMW is its clunky gearchange. The new design gearbox has left this tradition far behind. You can poke the lever through the box as quick as you like on upward changes with no worries whatever, but a degree of patience is still needed when changing down again. The downward changes can be made with no protest from the gearbox once the correct drill is learned — and quickly too — but you *must* use the clutch at all times.

If you want to ride the R90S like an Oriental super-screamer there is nothing to stop you. Its acceleration is very useful indeed if its full potential is unleashed. But don't be too dismayed to find the fleeter of the opposition's heavyweights hosing off the German twin without too much trouble, because it doesn't really shine in that department.

All you need to do is wait for the first set of bumpy curves to come rushing up to you and the boot is very much on the other foot. Here is one department in which the BMW *does* shine. The bike can be cranked over to some of the most improbable angles with little regard for road surfaces, and it will stick to a chosen line like lipstick to a singlet. It has got to be one of the few machines which can be flung about almost willy-nilly; your line through a corner altered at will as you lift the bike slightly to blast round the outside of your quarry, or drop it over further to dive underneath. So much for your torque reaction.

Under brakes the bike's directional stability is exceptional. The long, slow-travel forks and excellent three-way adjustable rear units soak up bumps almost as though they aren't there, but the occasional deep dip can catch you unawares. It happened once and the bike seemed as though it wanted to wallow, but the moment the bike was dropped into the corner it settled down. It left the heightening impression that such a bump was more unsettling to the *rider* than the machine itself, and that it was *he* who had to settle down to the job.

One odd impression is generated by the overall line of the machine. The starter motor and air filter box are located within castings on top of the crankcase, the huge mass of alloy filling the entire area under the tank and apparently leaving no daylight between engine and frame tubes. This, allied to the tall, thin appearance of the rest of the bike and with the cylinders sticking out like a diner's elbows, gives one the odd feeling of sitting on top of a pyramid. The impression belongs only to the R90S, for a quick burst on the other 900 cm^3 variant, the R90/6, dispelled the feeling entirely.

Saddle height is high and the bars are flat, which adds to the sports appearance of the bike and certainly helps one to tuck in behind the fairing when things start to get busy at the top end of the machine's performance (just over the metric double ton incidentally). The seat is commendably wide and a little firmer than the earlier models adopted. The neatly streamlined tail contains a tiny "boot" in which the prudent owner may care to carry some small items of food to sustain him on the long trips he is bound to take on this easy road-burner.

In top gear the bike cruises perfectly happily at 120 km/h with just 4400 rpm on the clock, but it raps along just as happily half as quickly again — it just gets there sooner!

The exhaust is routed through a small coupling pipe which joins both outlets under the front of the crankcases, and it discharges through a pair of long, upswept slow-taper mufflers which burst ostentatiously from each side of the machine's rear end. They are, of course, perfectly chromed and emit nothing more than an apologetic mutter at all stages of the bike's performance range. They will disappoint those who wish to announce their rapid approach to all within a three-mile radius, but are to be commended for displaying how a machine can be at once a rapid projectile and yet be discreet about it.

The test machine was finished in the optional Daytona Orange, and was good-looking even when covered in a thin film of dust after one particularly long, hard squirt. It covered the long trip with ease and was totally free of any drama. In fact its performance was almost monotonous, for it did everything expected of it with no surprises.

The only blot on the bike's copy book could be the front edge of the exhaust pipes, which were a beautiful blue tapering to straw in the tradition of almost every bike that has gone before.

There were no oil leaks at engine, gearbox or drive shaft joints, which is as it should be. It's comforting to know that the totally enclosed final-drive shaft allows maximum transmission lubrication without bathing the rear of the machine in oil or its rider's coat in little black spots. The Cardan shaft is carried inside the right tube of the swing arm fork where it can be ignored, except for 10,000 km oil changes.

Oh, and what about the little holes in the side of the rear hub that let you check brake-shoe wear at a glance with the wheel still in the frame; and the fully adjustable pillion footrests? Think about the advantages of adjustable, tapered roller bearings for the swing arm pivot, and what about tank depressions for your knees to keep you in place under brakes and the totally logical fuel taps you can switch on in the dark without having to find where on and reserve are? How do you feel about . . .

It is only a natural reticence, tinged with a feeling for the fitness of things, which stops TWO WHEELS raving insufferably about the BMW sportster for it is very hard to fault and very easy to ride.

It isn't perfect and it costs big money to put it on the road, but in this life you get only what you are willing to pay for. Or can afford. *

BMW R75/6

Touring suitability has been the 750's reputation for years.
Performance never entered the picture.
Now it has both.

• Touring machines are the most complex, sophisticated and expensive in the sport. The Japanese have run their computers beyond valve-float for the last half-decade developing bigger, faster and smoother road-burners. At the same time the Europeans have been honing, polishing and enlarging life-long designs. The result of these efforts is an array of open-road machinery beyond the comprehension of yesteryear's touring rider.

The recent past has decorated the present with handfuls of four-cylinder air-cooled four-strokes, liquid-cooled four-cylinder four-strokes, liquid-cooled two-stroke triples and single-rotor Wankels. It has produced air-cooled two-strokes in two- and three-cylinder configurations, ninety-degree four-stroke twins with overhead camshafts, four-stroke triples with pushrods. Science has fairly rained on the touring rider—yet the BMW endures. That it has endured for a half-century, and thrives today in the face of outrageous technological pressure applied by the Japanese, speaks volumes about the German firm's concern for function. Comfort in a touring bike is the essence of function. BMW has always cared about comfort—more than performance, more than engineering gimmickry, more than price-sensitive design—and it is this precise consideration that disengages BMWs from all other road motorcycles.

Most manufacturers, particularly the Japanese, produce separate engine, chassis and suspension components for their varied range of road bikes. BMW does not. The 600cc, 750cc and two 900s use the same chassis, suspension units and engine configurations. Excluding the Sport's trim, all the BMWs are mates with engine displacement and gearing the only variances. Therefore the R75/6 must be thought of as a BMW with more power than the R60/6 and less than the R90/6. There are no other differences.

The BMW fork is the most efficient in the touring field—regardless of machine size. Like the Maico fork, which is lauded as the best in motocross, the BMW unit is physically large and simple in design. The 36mm fork stanchions contain extra-long internal springs. Mated with lengthy aluminum sliders, the fork delivers eight inches of travel. This amount of travel, twice as much as many big tourers, allows the use of soft springs that compress two inches under the weight of the unladen bike. Even with this natural preload, the fork still has more working travel than other tourers.

The secret to the BMW's comfort and supple ride comes from more than just soft springs. The fit of the damper rod, damping control washers and springs, due to abundant clearances, produces a minimal of material friction. Friction has been further controlled by undercutting a section of the slider bore where the stanchion is supported. This undercut section reduces the frictional drag from the damping oil compressed between the two supports.

Cycle Test

The full-length bore of the slider makes for much closer working tolerances. The clearance between the slider bore and stanchion is far more consistent than with sliders that are bored into but not all the way through. Only the BMW and Maico use this fork manufacturing process. And only these two fork units work with proper smoothness and lack of frictional drag.

With the extra-long and largely stiction-free fork capable of absorbing most road undulations, the BMW's rear shocks are not overworked. Like the fork, the Boge shocks are softly sprung and gently damped. The Boges allow the rear wheel to move five inches. Only recent versions of LTS (Long Travel Suspension) motocrossers can match or exceed the BMW's suspension travel lengths.

While it is generally assumed that the heavier the machine the better the ride, the BMW is the lightest of all 750cc-or-over tourers. Conversely, the BMW has the highest Gross Vehicle Weight Rating of any road machine except the Harley Big Twin—which means the BMW can safely carry more weight in passenger poundage and luggage.

Comfort, the BMW's forte, is further enhanced by its saddle. The seat is no

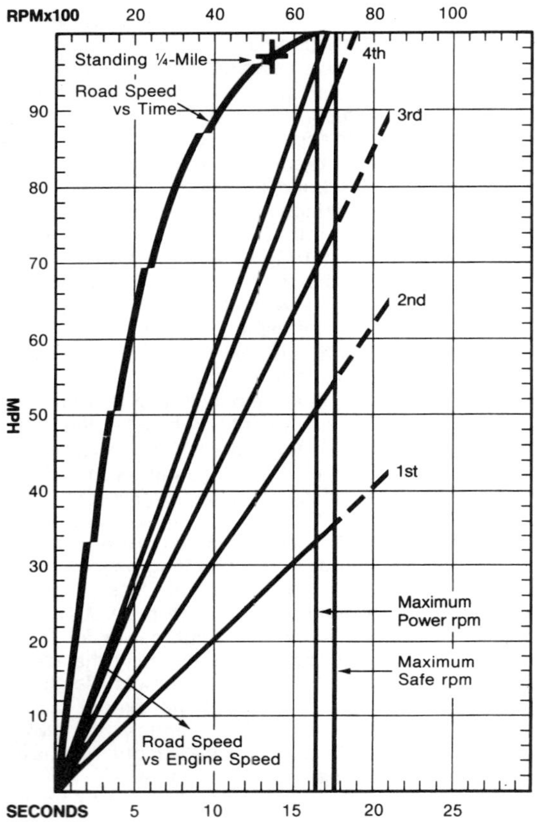

BMW R75/6

Price, suggested retail	$3195
Tire, front	3.25 x 19 Metzeler
rear	4.00 x 18 Metzeler
Brake, front	1.375 x 10.24 in. x 2 (35 x 260mm x 2)
rear	7.87 x 1.18 in. (200 x 30mm)
Brake swept area	105.5 sq. in. (680.5 sq. cm)
Specific brake loading	6.2 lbs./sq. in. at test weight
Engine type	OHV opposed twin
Bore and stroke	82 x 70.6mm
Piston displacement	745cc
Compression ratio	9.0:1
Carburetion	2; 32mm; Bing CV
Air filtration	Dry paper
Ignition	Battery and coil
Mph/1000 rpm, top gear	15.5 mph
Fuel capacity	4.8 gal. (18 liters)
Oil capacity	4.7 pts. (2.25 liters)
Transmission oil capacity	1.7 pts. (.8 liters)
Electrical power	280 watt alternator
Battery	12V, 25AH
Primary transmission	Helical gear
Secondary transmission	Driveshaft, crown wheel and pinion, 3.36:1
Gear ratios, overall	(1) 14.08 (2) 9.15 (3) 6.24 (4) 5.34 (5) 4.80
Wheelbase	57.5 in. (146.5 cm)
Seat height	30 in. (76.2 cm)
Ground clearance	7 in. (17.8 cm)
Curb weight	480 lbs. (217.7 kg)
Test weight	640 lbs. (290.3 kg)
Instruments	Tachometer, speedometer, tripmeter
Standing start ¼-mile	13.516 sec.; 96.77 mph
Average fuel consumption	49 mpg
Speedometer error	30 mph actual 26.92 60 mph actual 53.85

longer or thicker than other saddles, yet it is by far the most comfortable. The pilot's section is ideally recessed and shaped to conform to any rider's buttocks. The passenger section is elevated with a mid-way kick-up separating the pilot and passenger sections. The padding, though not excessively thick or firm, keeps rider and passenger posteriors from finding rock-bottom after day-long rides. The finish of the vinyl cover is serrated and semi-porous to keep undesirable slipping and sliding to a minimum. For the passenger there is a full 180° hand-rail bar that provides additional security and comfort.

The use of common parts for all four BMW models provides an enormous amount of interchangeability. The crankshaft, identical in all models, is a one-piece forging which rides in plain main bearings. The rod throws are 180° apart. Engine castings are high-pressure-formed aluminum with a sandy rather than polished or smooth finish, which aids cooling and reduces sound resonance. Lubrication is *via* an Eaton wet-sump system. The gearbox resides in a separate oil reservoir.

Huge variances in crankcase pressures are inherent with the pistons driving in and out simultaneously. These are compensated by a vacuum/pressure breather system that vents into the carburetion intake chamber. The chamber is located in the upper section of the engine's main tunnel housing (which also contains the air cleaner and starter motor), and silences both intake drone and crankcase pressure pulsations.

The camshaft is located under the crankshaft and is driven by a timing chain at half-speed. The ignition points and mechanical advance are located on the camshaft's end. Removal of the cast aluminum front tunnel housing cover will expose the ignition system and crankshaft driven alternator.

The valve train, simple by today's standards, is composed of old-fashioned pushrods, rocker arms and solid lifters. Located on the underside of the cylinders, the pushrods are external and enclosed in tubular covers. These pushrods are exceptionally long and relate to the engine's nominal 7000 rpm redline: overrevving will bend them in a hurry.

Power is transmitted to the five-speed gearbox through a single-plate dry clutch mounted on a large, automotive-type flywheel. The gearbox, turning at reduced speed, is located in line with the crankshaft. Power from the input shaft drives to the mainshaft into the layshaft; the rear of the layshaft, or output shaft, fits to the driveshaft; one universal joint fits between the output shaft and driveshaft, enclosed in a flexible rubber boot attached between the gearbox and right swing arm and shaft enclosure tube.

Inside the swing arm/shaft enclosure is a splined coupler that fits on the driveshaft's end. The shaft can move back and forth on the coupler and adjust for length variances as the swing arm moves up and down. The rear drive unit is, basically, an automotive-type ring-and-pinion assembly. BMW has six rear drive ratios ranging from 3.56:1 to 2.91:1. These have come from the 500cc through 900cc models, and all are interchangeable.

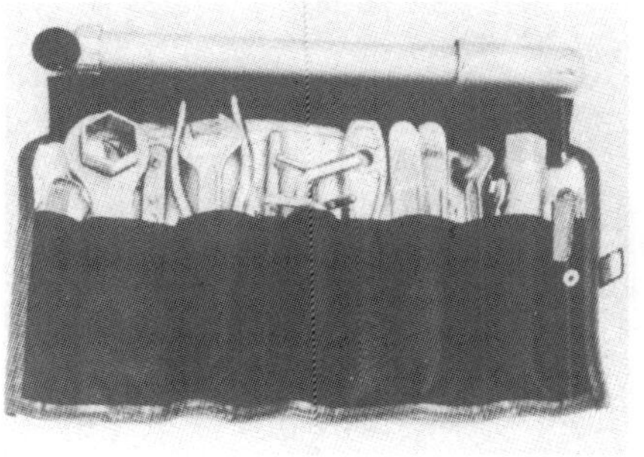

The best tool selection in the business includes a tire pump which is stored under the locking seat.

BMW has a neat cam-and-chain throttle but the blinker should operate horizontally from the left.

PHOTOGRAPHY: DALE BOLLER

No bike has a better reputation for two-up touring comfort. A soft seat and no vibration are the secrets.

BMW has traditionally claimed a small, hard-core segment of the street and touring market. In an effort to expand their potential, BMW has made a number of major production changes during the past two years. To improve handling the swing arm was lengthened; to broaden versatility a new five-speed transmission replaced the four-speeder; a disc brake took the place of the sluggish drum binder; new carburetors, higher compression ratios and more efficient mufflers improved engine performance; and the 900cc models, including the R90S hot sport version, finalized major shifts in BMW marketing trends.

The R75/6 is a surprisingly quick road machine. Its quarter-mile figures of 13.51 seconds and 96.77 mph are two tenths quicker and two mph faster than those of the R90/6. Much of the 750's impressive standing-start acceleration is due to a new, lower final drive ratio. Increased touring accessories and passenger-load weights, and the slower 55 mph speed limit, mean that the engine of the R75/6 will be less-burdened if allowed to spin

Range from the hand-striped silver-blue tank is about 240 miles. The Bing carb is West German.

Spider hub-design uses straight-pull spokes which are stronger than ones which bend at the flange.

Rear-mounted caliper keeps mass in back of the axle for lighter steering. Holes punched in the disc help reduce unsprung weight.

BMW R75/6

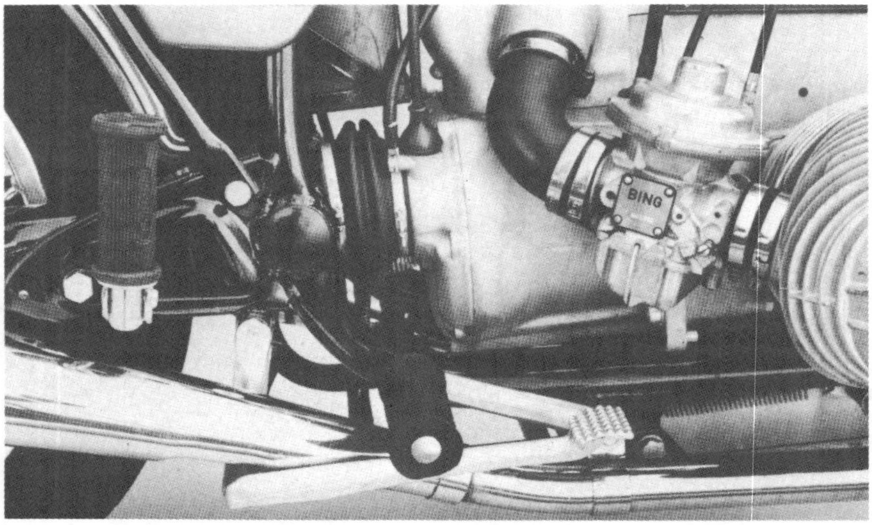

Inside the pleated rubber boot above the footpeg is an automotive-style U-joint which joins the transmission output shaft with the driveshaft, which is located inside the swing-arm. U-joint retail is $98.60.

faster. The gearing change from 3.2 to 3.36 has not resulted in a large jump in engine speed; at 60 mph, the engine now turns 4064 rpm instead of 3870.

On the road, the R75/6 produces more impressive results in gear-to-gear roll-ons with other 750s. In top gear twist-ups against a freshly-tuned Ducati GT 750, the R75/6 comfortably walked away from the Italian V-twin. Against the Honda 750 four-piper or Suzuki triple the BMW stretches out an even wider gap.

Traditional BMW lightness and this year's shorter gearing make the 750 more responsive than before. Slow-traffic poking is simplified as the engine runs happier, and with less torque reaction, in the lower three gears. Handling is more than acceptable for long-distance commuting and open-road touring. The R75/6 runs true, with no tendencies to twitch or oscillate up to the bike's top speed. Most riders will not object to the BMW's characteristics on twisty mountain roads. Steering is predictable and, up to its ground clearance and tire limitations, the BMW performs adaquately on bending mountain roads.

The BMW's handling restrictions come from the flat-twin cylinder protrusions, supple suspension units and hard, low-profile tires. In sharp off-speed corners and fast (particularly bumpy) turns the BMW's right valve cover and center-stand tab can be brushed onto the pavement. The Boge shock dampers will fade when the BMW is ridden hard, and cause it to wallow even in smooth turns.

In the ergonomics department (ease with which a rider and passenger fit and operate the bike) the BMW mixes excellent and mediocre qualities. Comfort relating to suspension compliance, seating position(s) and vibration control are, combined, the best in touring—solo or two-up. But other controls and BMW-inherent hindrances stand in the way of perfection. One annoyance that a BMW rider must live with is the constant contact between shin-bones and the air intake tubes from the carburetors. Hooks on lace-up boots catch on the tubes or carburetors. The rear brake lever arcs to near-excess when applied. Even slight rear brake free-play lets the lever carry the rider's toes too close to the ground in turns. The handlebar on the R75/6 is swept back sharply, causing the rider to bend his wrists awkwardly.

The throttle effort required is not acceptable. A combination of friction from the cam-gear-chain-driven twist grip, drag

in the throttle cable and stiff return springs aggravate the right wrist and forearm. Softer springs and well-lubed or nylon-type cables would greatly alleviate the problem. The clutch has a narrow friction point and requires a moment's thought before releasing the lever.

More discomforting are the electrical switches. They were designed specifically by BMW for federal regulation compliance—but unfortunately they were designed by automotive standards. The turn signal switch, located on the right (throttle) side, flips down for left and up for right—a directional control unnatural for a motorcycle rider. They should move side-to-side—right for right turns, etc. The light control for high, low and dip is on the left side, and also works in an up-and-down fashion.

The BMW's instruments are militaristic in simple, black-background design. They are, however, easy to read and glare-free. The speedometer is 10% fast while the odometer and tripmeter indicate mileage perfectly. BMW has the best headlight in motorcycling. The dual-filament quartz halogen light has a half-moon-shaped low beam and a high beam which are both much brighter than conventional sealed-beam lamps.

Lower final drive gearing has not affected the BMW's gas mileage. Riding two-up (and quite hard) we averaged 43 mpg, and watching the speed limits we saw 55 mpg when solo. An optional 5.8 gallon touring tank is available for long distance riders for an additional $60. Alternative final drive gearing set-ups are likewise available, but must be arranged through the dealer.

Despite scintillating engine performance that seems somewhat out of character, the BMW's hallmark is comfort—comfort over distances that riders of other brands would find more punishing than amusing. The suspension settings, saddle design and vibration control all contribute to a state of mind that is relaxed and serene. The R75/6 is composed in everything it does except hurtle through the mountains—and BMW figures if that's what you want, then you should do it aboard something else. The bike knows what it is, and what it can do—and makes no pretense about conquering other venues. Being the best 750cc tourer, and one of the best tourers in any displacement class, is more than enough for the BMW R75/6. ◉

CYCLECAMPING

Continued from page 51

instrument panel lamp.

We encountered only minor mechanical failure—on our return trip the speedometer broke. The tach worked out well in gauging speed for over 2000 miles.

It wasn't easy to fit all the equipment on the bikes. Both saddlebags were meticulously packed. Bulkier things—sleeping bags, ground insulating pads, nesting utensil kit, along with a few clothes, were packed in a nylon duffle bag. This was secured over the tent with tension cords that tied into four eye bolts on a 3/4-in. board mounted to the luggage rack.

The fairing was found to be worth its weight in gold. It not only gave protection from the wind (resulting in less driving fatigue), but also provided valuable storage space for our nylon rainsuits, the spare inner tube, and maps. The spare parts were taped securely inside the headlight.

Trial trips—We took two trial trips prior to the long, cross-country haul. It was well worth it. The experience of the shorter trips. gave us a trial-and-error system of proving not only ourselves but also the equipment.

Campsites—At times finding the right campsite was frustrating, but we learned. Knowing what was up the road the next day made the next evening more enjoyable. Talking to fellow campers who were traveling in the opposite direction gave us some indication of what we might expect.

The AAA's western edition of Camping & Trailering didn't prove too helpful as a guide to finding adequate campsites. It appears that most guides list every cow pasture that farmers find no longer profitable for cattle. The old barn is usually headquarters for the "dude ranch." The fireplaces at each campsite turns out to be a pile of field stones laid in a circle.

What the guides don't list is the noise factor of the approved camp areas. Our experience in one area near Salt Lake City, Utah, was horrible. The tent area was located close to a freeway, railroad tracks, and an all-night trucking terminal. The noise was unbearable—but worse, the ground even vibrated from the truck and train traffic! Pick your campsite carefully!

Planning your day is important. We found it beneficial to find a campsite early—say around 4:00. We were more often assured of a better site before the mass of travelers put in for the night. Breaking camp early and getting on the road as soon as possible was at times hard on the tent when the dew had not dried. But, it was worth the extra effort, especially in taking advantage of cooler mornings. ◘

Sahara

Continued from page 119

thing. Turning around, the whole world changed. Tire tracks went in all directions muddled with camel prints. What should have been an easy task became a nightmare in a hard patch of gravel where we lost our treads. We doubled back and spread out, only to confuse ourselves with our own tracks.

"What now?"

"We must stick with the bulk of the tracks and find the well again," I reasoned.

"Which dozen do you prefer?"

"North. Follow the compass north."

The camel tracks gave us a clue and eventually we sighted the ruined building. Scanning the horizon, something caught Fran's eye to the East. It was not a post. It looked more like a tombstone. Probably some tourist's grave, I imagined. We rode toward it.

The tire tracks became more numerous as we came up to a cement marker shaped like the hundred pound weight of a circus strongman. The highway department had switched markers on us at the only turn in 1500 miles. We had found the track: east by southeast.

Later in camp, relief gave the vegetable beef soup an extra flavor as we pretended not to notice having broken into our reserve survival rations. The last of our gas went into half empty tanks, and only the sulfurous Assamaka water was left to drink. The track was a road again and somewhere at the end was Agadez. How far was the question.

In the morning we took bets on the mileage to Agadez. Nobody put their money on less than a hundred miles. Surrounding us was the Sahel where the drought had strangled the nomads' cattle, driving the survivors to the relief centers. The thorn trees had taken over. The corrugations and sand holes soon forced us off the road, dancing between the bushes where a Rover could not go.

Coming up a rise, we saw it. There on a hill was the red and white striped water tower of Agadez. We did it! We crossed the Sahara! Riding three abreast we roared into the sleepy town all smiles and running on reserve. There on the main street was Shasti.

"Where have you been?" she asked. "We expected you yesterday."

"Wanted to enjoy the scenery," winked Fran.

There was more desert to the south, but we were veterans. It could not be worse than Assamaka.

Two days later at home in America, our parents drank a relieved toast as they read the telegram: SAHARA CONQUERED HAPPY NEW YEAR YOUR SONS. ◘

LLE
513
P

Giant Test

In Search Of Ultimates

BMW R90S v Moto Guzzi S3

Testers: Bill Haylock, Mike Nicks.

IF AN ENTERPRISING dealer assembled a random selection of 73 motorcycles in capacity classes from 50 to 1,200 cc and prices from £195 to £2,500, his showroom would surely have enough variety to satisfy every nit-picking whim. Or would it? That figure of 73 happens to be the number of bikes that has passed through our hands for road test or impression in the 15 issues up to this one that we've compiled since *Bike* moved to monthly publishing schedules last summer.

Yet still there was something missing. We had yet to find the ultimate long distance motorcycle; a machine that could hold a level 100 mph without mechanical hysterics, that could carry enough fuel to keep its motor beating for at least 200 miles without break, that would give the rider full tilt highway pleasure for hours at a time without racking his limbs with muscle fatigue.

Many of what are optimistically called superbikes ought to have these features but don't. Maybe the handling or the high speed stability is suspect, the engine a shaker, the fuel consumption too much for a puny tank — or maybe the thing's just too uncomfortable to ride fast for very long, which is the most frequent complaint with modern bikes.

In fact when we got down to compiling a list of runners for this particular Giant Test we came up with just two serious contenders: BMW's R90S and Moto Guzzi's 750 cc S3. Now if your eyebrows are raised and your mouth agape because your own favourite Supadupabike didn't make it in our estimation, we'll ask you to ponder awhile on whether your ideal steed really would meet the criteria we'd laid down. By 100 mph cruising, for instance, we mean the real thing and not a flash reading taken off a grossly optimistic speedo while nuts and bolts vibrate merrily off the bike in protest at the madness of it all.

Anyway, the 90S was an obvious choice, a known and trusted design of impeccable heritage. But a few hoary old myths were still waiting to be laid: would the torque reaction from the in-line crank hurl us off if we hit so much as a cat's eye at 90 mph? Had the once notorious BMW gear change really been smoothed out?

The Guzzi was more of a gamble. Barnaby Williams had written glowingly about it earlier in the year, but now there was a revised model with at least one radically new feature, more of which later. The Guzzi is also a mere 750, whereas we'd already left off the list a certain widely respected 900 (sorry Mr K, the performance is fine but for the unusual demands of this test high speed rider comfort was doubtful). However, the importers showed no qualms when they knew we were going to match the S3 against the BMW flagship, so we pressed on regardless.

Photography: Mike English

BMW R90S

EVERY time I commit my feelings about a certain bike to paper, I find myself tempering my opinions. It's a question of compromising between my unfettered opinion on the intrinsic good or bad points of the bike on one hand, and what the limitations imposed by my finances and needs would allow me to think if I was in the position of the Average Customer nervously approaching the dealer's showroom.

Inevitably having thus tempered my judgement, Annoyed of Neasden *still* writes an irate letter to the effect that "You idle buggers swanning around on MVs and Benelli Sixes all day haven't the faintest idea what us ordinary working class bikers want to know". Such vitriolic attacks wouldn't sting so much if they were true, but sadly most of our miles are covered on much more mundane transport — I mean, *you* don't have to ride sports mopeds if you don't want to, do you?

So it's a mighty relief when at last I come across a bike that makes me feel like saying "Sod everyone — this is a bike I've not got to make any excuses for" . . . A bike that stands on its own merits without having to look at it through the eyes of the middleaged, penny pinching commuter, or the sixteener-banned mopedist, or the posturing cafe racer.

The fact that such a bike should be a very expensive quality product, is inevitable, I suppose. And that doesn't help to placate Annoyed of Neasden at all, when he was wondering if he could afford the meagre roubles required to become the proud(?) owner of a certain antiquated, but extremely cheap Iron Curtain bolide.

BMW's flagship, the R90S, may not be the ultimate, but it's probably the nearest anyone's got to it yet. And Mr Annoyed, it's pointless you bleating: "But can a motorcycle be worth 2,000 quid?" If you've got the bread you too can have all the discerning motorcyclist needs to ride as safely and fast and comfortably as possible — well as much as money can buy off the showroom floor. No matter if you've got to beg, borrow, rob or put yourself in hock to get the loot, it's worth it.

When the R90S was first introduced LJK Setright described it as ". . . possibly the best production motorcycle in the world . . ." Two years further on we've become more blasé about bikes costing two grand now quite a handful of manufacturers have prestige machinery in that league. But I believe the BMW is still the best.

Now I'm in danger of sounding like any other impressionable road tester, about to get stuck up to my neck in a swamp of sickly superlatives. Before I disappear from sight I'll attempt to justify such a sweeping statement.

The R90S handles and stops almost as well as the best Italian sportster; is almost as fast as the fastest Japanese roadburner; it is almost as uncomplicated as the good ole British Twin; it is almost as smooth as the best multi; and when it comes to comfort and capability for travelling at maximum speed with minimum fatigue, the R90S is second to none.

Razor sharp as you all are, you will no doubt have noticed that in most respects I only claim it to be second best. The truth is that many bikes do one thing superbly, but the BMW is the only bike I know that does *everything* very well.

The reason for that is the Bayerishe Motoren Werke's uncompromising approach to the design of a motorcycle. They obviously sat back and thought very hard about the fundamental functions of a motorcycle before they put pencil to paper and, what's more, you get the impression they approached it from a biker's point of view, rather than an engineer's. They thought first of all it should have a light,

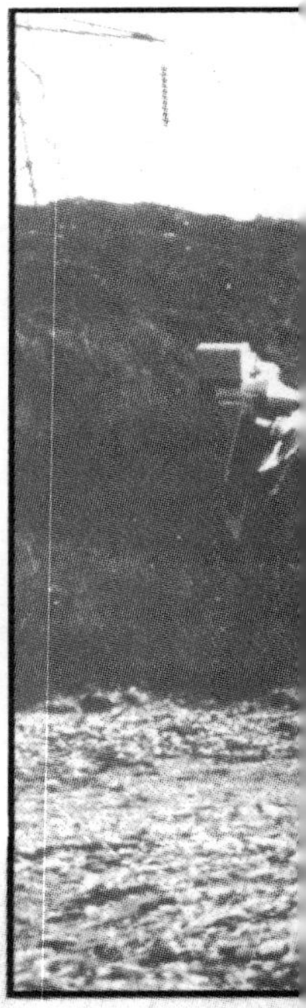

750-S3

bike

simple, powerful, low maintenance power/transmission unit. They already had that in the transverse flat twin layout they'd been refining since the twenties, so they refined it a little further and made it more powerful.

They decided it should be manageable in city traffic, and yet be steady as a rock at its maximum speed and capable of cornering to the limit of tyre adhesion. So they

designed one of the best balanced chassis/suspension layouts yet produced.

They thought the rider ought to be able to cover 200 miles of fast travelling without having to stop. So they made the controls as easy to operate as possible, the riding position as unfatiguing as possible, and the petrol tank large enough to carry ample fuel for that distance.

Checkout

	MOTO GUZZI 750 S3	BMW R90S
Engine	ohv 90-degree V-twin	ohv flat twin
Bore and stroke	82.5 x 70 mm	90 x 70.6 mm
Capacity	748 cc	898 cc
Compression ratio	9.8:1	9.5:1
Carburation	2 x 30 mm Dell 'Orto	2 x 38 mm Dell 'Orto
BHP @ RPM	72 @ 7,000	67 @ 7,000
Primary drive	Direct	Direct
Clutch	Twin-plate, dry	Single-plate, dry
Gear ratios		
(1)	10.806:1	13.2:1
(2)	7.499:1	8.58:1
(3)	5.657:1	6.21:1
(4)	4.695:1	5.0:1
(5)	4.052:1	4.5:1
Electrical system	12v generator, battery and coil ignition	12v 280 watt generator, battery and coil ignition.
Lighting	45/40w headlight, 2/20w tail/stoplight	60/65w headlight, 5/21w tail/stoplight
DIMENSIONS		
Wheelbase	58.75 ins	57.7 ins
Seat height	28.75 ins	32 ins
Ground clearance	6.75 ins	7.0 ins
Handlebar height	32 ins	—
Kerb weight (with 1 gal fuel)	500 lbs	466 lbs
Fuel capacity	4·8 gals	5.25 gals
Oil capacity	6 pints	4 pints
EQUIPMENT		
Trafficators	Yes	Yes
Electric starter	Yes	Yes
Trip mileometer	No	Yes
Steering lock	Yes	Yes
Helmet lock	No	Yes
Headlight flasher	Yes	Yes
Kill button	Yes	Yes
Toolkit	Yes	Yes
Spare parts	No	No
Others	Shaft drive, alloy rims. coupled braking system, steering damper.	Shaft drive, alloy rims, steering damper, clock, voltmeter, tyre pump, mirrors, fairing.
CYCLE PARTS		
Tyres (front)	3.25 x 18 Michelin	3.25 x 19 Continental
(rear)	3.50 x 18 Michelin	4.00 x 18 Continental
Brakes (front)	Twin 11.8 in discs	Twin 10.24 in discs
(rear)	Single 11.8 in disc	7.9 in sls drum
PERFORMANCE		
Top speed	123.45 mph	126.06 mph
Standing ¼ mile	14.60 secs	13.23 secs
0-60 mph	6.38 secs	5.19 secs
Speedometer error		
at indicated 30 mph	23.75 mph	27.7 mph
at indicated 60 mph	50.43 mph	57.14 mph
Braking distance		
from 30 mph	26 ft 6 ins	32 ft
from 60 mph	137 ft 6 ins	156 ft
Fuel consumption		
overall	45 mpg	43.7 mpg
ridden hard	41 mpg	38 mpg
PRICE	£1,749 inc VAT	£1,997 inc VAT
GUARANTEE	4,000 miles/6 months parts and labour	12 months unlimited mileage, parts and labour
Supplied by	Moto Guzzi Concessionaires, Crawley Road, Luton, Beds.	BMW Concessionaires, 991 Great West Road, Brentford, Middlesex.

BMW R90S

That set of prerequisites happily coincides with my idea of what motorcycles are all about, and produces what I believe is the fastest, safest and most comfortable means of travelling on two wheels.

What is missing perhaps is the element of extrovert style that the Italians and Japanese are so good at presenting. The R90S is no bike for the techno-freaks, its essence is that of quietly understated quality. Maybe it's a sign I'm getting old when I start appreciating things like that — or should I say more mature, sounds better. Obviously the BMW is aimed at older riders 'cos they're the ones with the bread and there aren't enough 19-year-old rock stars around to keep BMW going.

Certain touches remind you of the rising young exec BMW's marketing men seem to have in mind as the Average Customer (or should it be Client?) The clock for example, no doubt indispensible to the said young exec dashing between appointments in the City. Actually it's no mere gimmick — it's nice to know how the time is going on a long trip and it's a bloody nuisance to have to stop to check your wrist watch. Typical of the BMW's Teutonic efficiency, the clock seems more accurate than most watches too.

To see the BMW and Guzzi side by side is revealing about the national characteristics of the two countries that produced these bikes. The BMW looks awkward and stolid beside the Guzzi's sleek beauty which is the ultimate in the sort of tasteful, sensuous styling only Italians can produce. But in close-up things can be seen in a different light. The Guzzi unfortunately suffers some of the failings characteristic of all Italian bikes, with its poorer finish and lack of thought to details. With that unnerving Teutonic thoroughness, BMW's designers seem to have thought of everything the Italians forgot.

It is very apparent that the whole of the R90S was designed as an entity, down to every last control lever and switch, and that's where the German factory scores over the Italians with their propensity for spoiling very fine ships for a halfpennyworth of tar, or cheap mass produced accessories.

Ergonomics is a word that most bike designers don't seem to understand, but typical of the thought that BMW have put into the ease of operating their bike is the unusual design of the switchgear. Redesigned since the bike's introduction, this now provides on each side of the bars, a rocker switch operating in the vertical plane, a push button above that, and a toggle switch away from the handle bar grips. On the left side of the bars the toggle switch operates the lights — off, pilot light and headlight — the rocker switch operates main and dip beams, and headlight flasher. The push button works the horn. On the right bar the toggle switch is a three position engine kill switch, the rocker switch operates the flashers and the push button is the start switch.

The vertical operation of the flasher switch felt unfamiliar at first, but then I realised that all the switches you need on the move — indicators, horn, headlight flasher and dipswitch, can be reached by merely rotating the thumb of either hand through an arc of 90 degrees, without changing the hand's grip on the bars.

BMW's designers seem to have struck a magical compromise in producing a riding position that suits people of quite diverse shapes and sizes. My gangling six foot frame tucks in neatly behind the narrow bars and sleek handlebar fairing. Yet, funnily enough, the shortarses in this office who usually complain about the riding position on bikes that fit me perfectly well, were for once in agreement with me that the 90S is the most comfortable bike around when it comes to enduring long hours of hard travelling.

Flat, narrow bars contribute a lot to the tidy, tucked in feel of the riding position, and the footpegs are set well enough back to provide that well balanced, slightly forward leaning stance which produces no aches or fatigue after two hundred miles or more. The seat is wide and well padded, and the narrow tank helps keep your knees tucked in.

The small fairing which is part of the stock specification on the 90S is not just cosmetic. Despite its meagre proportions, the superb quality glasswork keeps a surprising amount of wind off your chest and head, making 100 mph cruising perfectly practical, apart from the sinister proliferation of those spoilsports in white Rovers and Volvos on any decent stretch of tarmac.

The fairing also provides a mounting for the volt meter (indicates the battery's state of charge) and electric clock, presenting the rider with a neat and functional cockpit. As you'd expect, the instrumentation is excellent and unfussy, with night-time illumination bright enough to read the well-placed clocks at a glance, but not bright enough to distract or dazzle.

Control layout and riding position is greatly responsible for making the 90S such a fun bike. It's one of those machines you take a pleasure in handling really well, so responsive is it to the owner's control, the big BMW inspires a confident and precise riding style.

You'll always find people trying to tell you horrific tales of cylinder heads ploughing the tarmac through bends, and the fearful effects of torque reaction (due to the crankshaft spinning in the same plane as the direction of travel) on handling. It's crap, probably inspired by envy. The truth is the 90S is one of the best handling big bikes around.

Part of the reason for this is its creditable *avoir dupois,* being a mere lightweight at 466 lbs compared with the gross behemoths the Japs have taken to creating. Another reason is, that like everything else, the suspension is perfectly tuned to the rest of the machine so that the ride is both supremely stable and comfortable.

Stability is often achieved at the cost of comfort, but BMW have not sacrificed the bump absorbing qualities of the long travel suspension. They have also managed to produce a chassis geometry to give steering which is light and positive at low speed and yet combines hairline accuracy and total stability at maximum speed.

The only time handling can get less than reassuring is if you make clumsy gear changes going round corners. Any shaft drive bike is the same, because the shaft does not absorb transmission shocks like a chain does, and if you don't match revs and road speed you can momentarily lock the rear wheel when changing down. You've got to be pretty lead footed to do it though.

This is turning into an embarrassingly long list of superlatives which no doubt smacks of payola (well BMW, anytime you wanna give me a free R90S, just call me . . .) But there's still more to come before the moans.

Electrics are the best, especially that brilliant Bosch quartz halogen headlight which is a match for most car candle power. The toolkit is of the same sort of quality as the bike itself and comprehensive. You get a tyre pump underneath the hinged seat too.

So to the grumbles, and a niggardly list of minor nit-pickings most of them seem, but the general high quality of the bike does seem to accentuate anything not quite up to par. Like the black paint peeling off the top fork yoke — a surprising blight on the Bee Em's otherwise high quality finish.

Try as I might, I couldn't get the steering lock to work. Could have been my clumsy fingers or numb brain, but you shouldn't need an engineering degree just to make the bike thief-proof.

Perhaps the worst point is the gearchange. BMW's have always been infamous for the jarring harshness of their shift, and they've done a lot to rectify it, in fact the gearbox is one of the few items to be modified since the bike's introduction. But by comparison to most Japanese boxes, it's still far from slick, especially between the lower gears.

The position of the oil orifice in the sump means you'll be lucky to get less than half your lubricant over the garage forecourt or the garden path, unless you have a specially shaped funnel handy.

Given my predeliction for Italian bikes which I admit quite unashamedly, I was surprised to find I enjoyed riding the BMW more than the Guzzi. I find the BMW's appearance unexciting, and I'm infatuated by the Guzzi's looks — if I had that kind of money I'd buy the Guzzi on its styling alone. It would take a lot of soul searching though, to decide whether I need the Guzzi's beauty or the BMW's perfection.

Bill Haylock

Moto Guzzi S3

THE ITALIAN motorcycle revival that's been commanding so much attention in the press of late is no mere conjecturing on the part of writers short of happy events to cover in the home industry. If final proof is needed of the Italians' virile and positive approach to motorcycle design, it's there in plenty with the impressive Benelli six we tested in September, and now with the Guzzi S3.

What I like about the Guzzi is that it's a machine without compromise. Making a transverse V-twin must be a task full of pitfalls, not so much from an engineering standpoint but because it's an unusual layout that must be made visually acceptable to the notoriously fickle biking public. You can imagine the Guzzi engineers, halfway through the design, looking again at some of the archetypal superbikes of the seventies, the Honda four, the Z1, and now the Benelli Sei, and breaking into cold sweats.

But they haven't flunked the job, tried to disguise that it's a big bold traverse V, or slapped too many tin badges everywhere in a frantic attempt to prove that it's just like all those multis after all. Instead they've gone ahead with an individual, undiluted line of thought on the machine, and come out with the most beautiful production motorcycle you can buy. That's my opinion anyway, and to judge from the comments the S3 aroused while it was with us, I'm far from being alone.

The dominant feature of the Guzzi is not so much the engine layout, with its two huge cylinders jutting from a web of intricate looking castings below the tank, but its lowness. At less than 29 inches, the seat height is around three inches less than on most 750 cc-plus motorcycles. They've done it by making intelligent use of opportunities offered by the transverse V arrangement. The top frame rails actually run in between the cylinders, and even the well of the fuel tank sits below the top of the cylinder heads. Having got the frame rails that low, they've then run them straight back so that the seat height is less than on most 125s, if you can believe that. The S3 is almost the only — maybe *the* only — large motorcycle on which a short rider can sit with his boot soles planted firmly on the ground. Compared to the previous V-twin we tested, the towering 860 Ducati in October, the Guzzi is like a Lear jet parked alongside a 747.

Top: Massive V motor looks indestructible, feels it on the road. Above: S3's pillion footrests are a token gesture, as the seat is basically a one-person affair.

All this is complemented by what seem like the two longest exhaust pipes you've ever seen, curling away from the cylinders and ending at a point level with the back of the rear tyre. They're black, as is the frame, the tank and the seat, the orange band swathed across the tank being a typically bold Guzzi gesture. The tank itself is another of those Italian masterpieces, holding plenty of fuel (4.8 gallons) yet looking as though it's been sculpted from a block of solid steel. Sitting astride the bike, your eyes are drawn down the path of the orange band until you meet the words *Moto Guzzi* on the grey alloy of the rocker box covers peeking from underneath the tank. Subtle, and very effective.

So we have this motorcycle that looks like it must be a 130 mph rocket (or 140 mph, as at least one dealer is claiming in his adverts). But does the image survive the experience? Mainly yes, in some areas no.

Starting with the heart of the matter, the engine, let's point out now that this is one of the surprisingly few motorcycles that will hold cruising speeds in the high nineties without feeling like it's going to scatter its internals over the tarmac. For various reasons, some of them not unconnected with cars that have flashing blue lights on their roofs, you may rarely travel at such extravagant rates (although on the S3 100 mph is far from being a frantic experience). The point is, if you have a machine of such stamina, lower cruising speeds naturally come that much easier.

But while the Guzzi is an effortless high speed cruiser, it will fall short of many people's expectations because it doesn't accelerate in the manner to which we've become accustomed in our latterday "superbikes". Oddly, for a big-inch four-stroke of only two cylinders, most of the S3's claimed 72 horsepower is stacked in quite a narrow power band at the top of the rev range. Even when you buzz the motor above five grand the bike doesn't seem to lunge forward with the brute force of other biggies. The bike is also highly geared, and this, combined with the shortage of low speed torque, results in the distinctly moderate standing quarter time of 14.60 seconds.

However, once the initial disappointment is over you realise that things could hardly be otherwise. If you want rubber smoking starts *and* high roller cruising, you'll need 900 cc or more. And in any case, the S3 has the kind of balanced all-round performance that returns deceptively high average speeds.

This was illustrated during the test when Haylock and I launched westwards to take in the hill climb freakery in Devon mentioned in last month's issue. As we bid farewell to the romantic brickyards of Peterborough on a grey Saturday morning, a persistent rain drowned the human spirit. Aha, let's turn back, fire up the company car, and *motor* off for five hours of stereo-backed sensation, sealed off from the outside elements. Hey but look, we're supposed to be *pro testers,* hard man bikers, and the *Bike* credibility ratings are gonna crash if it gets around that we do too much mileage in red Cortinas! So the bikes it had to be — and then, as so often in these seemingly dead-end situations, fings turned out to be not so hopeless after all. It was quickly apparent

Giant Test

that the Guzzi was not one of those coronary - on - every - wet - corner frighteners. A combination of soft four-stroke power, fine handling from a strong frame and supple suspension, brakes with a high degree of feel, and that most crucial biker's item, good tyres, had the Guzzi gliding swiftly yet without drama along miles of wet and twisty Midland by-roads. The S3's reassuring rubber, incidentally, comes in the form of Michelin's S41 covers.

In the dry, as you might expect, riding the S3 is a rather fine experience. The motor has an odd wheezing sound that somehow fits the bike's loping style. Laying over the big tank, you aim the Guzzi into turns and it feels very predictable, very reassuring. As on all the best motorcycle gearboxes, you can sense the cogs moving out of and into mesh as you make the changes. It's a two part operation on the S3, with an almost imperceptible pause needed between the movements, yet it's still a smoother and much quieter business than on the BMW. If you're into the blurred scenery style of biking, you can't help but like the S3, and it's at such times that the £1,750 price ticket seems justified.

However, the Guzzi isn't the day-long full tilt land cruiser that we'd hoped for. The riding position, not the machine itself, makes that impossible. A medium height rider finds it quite a stretch over the tank, and despite the low seating position, the clip-on bars induce wrist and shoulder ache so that two hours is about as much riding as you want before you need to break. The seat is also too narrow for true long distance touring, and it's short enough to make two-up trips a distinctly intimate affair.

The equipment and fittings largely match the Guzzi's price. The refinement of shaft drive means running maintenance is little more than an oil check every 1,800 miles, and an oil change every 6,000, as opposed to desperate applications of aerosol lubricants to keep musclebike chains in order. The major change from the S3 featured in these pages earlier in the year is the coupled hydraulic braking system. The foot pedal operates the rear disc and one of the twin front discs, a compensator ensuring that 75 per cent of the pressure is applied to the front stopper. The result is highly predictable braking on even the greasiest surface, with the handlebar lever acting as reserve to bring in the other front disc for emergency stops. It's intelligent sophistication, unlike some of the suspect gimmickry being pressed upon motorcyclists largely by the Japanese makers.

Other useful safety features on the Guzzi are an electric starter that cannot be operated until the clutch lever is pulled in (to stop you from lurching forward if you forget you're in gear, dummies) and a catch button that prevents you from accidentally switching all the lights out. The switchgear itself is among the best to be found on Italian bikes, practical and easy to locate. A tool kit, excellent in both quantity and quality, is stored in a container below the seat, but really needs more room, and the centre stand is impossible for one person to operate in confidence. Well it was for the emaciated *Bike* staff, anyway.

Ah yes, nearly forgot that old fable about the dire effects of in-line crankshafts and shaft drives on motorcycles. It's almost entirely an illusion. If you rev the motor with the bike stationary, you can indeed feel it rocking slightly from side to side. But you don't notice it once you're moving. Failure to match revs to road speed when changing down can lock the rear wheel, possibly causing you to crash on wet roads or in a bend. But then the same sort of clumsy riding will send you down on any two-wheeler. Take it from us, the Guzzi is among the safest and most stable bikes you can buy.

Put through the speed trap with 1,500 miles on the clock, the S3 registered 123.45 mph. The importers say that Guzzi V motors are not fully run in until up to 4,000 miles have been covered, so maybe a little more speed was lurking inside the test bike. Personally, I doubt it, and in fact I feel obliged to shoot down some of the more fatuous performance claims that the S3 seems to inspire above all other motorcycles. Not only do we have the aforementioned 140 mph fantasy, but a recent road test on the machine claims that at 120 mph the engine is spinning at only 6,300 rpm, "well within its limits" and able to hold that speed "indefinitely". Maybe somebody hadn't been checking the Guzzi's erratic speedometer, because at 6,000 rpm in top it was actually running at 94.93 mph, according to our electronic timing gear. Flat out on the 123 mph run it was revving at just on seven grand.

In terms of general road use, however, even this information is slightly academic in that the S3's easy-rolling style is deceptively fast over most roads. Limitations in rider comfort prevented it from being the all-day mileage machine we were originally looking for in this test, but it's still one of the most appealing motorcycles in a very competitive section of the market.

Mike Nicks

Summary

WE WERE looking for ultimates, and our choice of machinery proved a good one. But although both bikes fulfilled nearly all the exacting requirements to qualify as the ideal long distance high speed cruisers, they reflect two quite different approaches to building bikes.

Both companies use very similar engine/transmission layouts. The transverse engines and shaft drive, providing tireless, robust power at sustained high speed with a minimum of maintenance, are ideally suited to the demands of the hard use these bikes are intended for.

But the way the design concept has been arrived at is very different. The German factory combines traditional high quality workmanship with typically painstaking thoroughness in applying logical thought to the machine. The Italians use not so much logic, as intuitive flair for motorcycle design, with an equally high engineering skill.

The thorough German approach results in a machine which is almost faultless and entirely functional. Meanwhile, the intuitive Italian approach produces a bike of highly individual character, with a few flaws. The Guzzi does not match the BMW on quality of detail or finish, and at £250 cheaper it's not fair to expect it to. Neither does it offer the same comfort as the R90S. But it makes up for it with beautiful styling and that indefinable quality character.

Which bike you choose in the end depends on what sort of person you are.

BMW R90/S

Breath-takingly priced, but also lavishly endowed with quality and the kind of performance that makes you wonder why everyone has to travel so slowly.

● As somebody once said, the law is a crude instrument. Here we are, obliged by inelastic law to travel at speeds no greater than 55 mph, and that is presumed to be without exception the right and moral thing to do. Lowered highway speeds, we are told, bring about a critically important reduction in the average American automobile's fuel consumption—if only down from the obscene to being merely grotesque. We also are told that this originally-economic legal stricture now has moral force because our traffic safety record is not quite so dismal at the lower limit, and that may even be true. Vehicles with a demonstrable reluctance to stop or turn collide with less frequency, or at least with less force, when driven at 55 than they did when streaking the Interstate highways at speeds up to 80 mph.

But then something will come along to demonstrate that Churchill had it right when he commented that "the law is an ass." For us, that demonstration is the new BMW R90S, which isn't decently in stride unless it's moving fast enough to give a highway patrolman apoplexy. And none of the rationalizations offered up by all the law makers and enforcers will keep the BMW from making the law itself look arbitrary and silly. They want to argue about energy conservation? The BMW can't be hammered hard enough to make it eat gasoline at a rate much worse than 45 mpg. Safety? Nothing capable of motion is entirely safe, but if they want to talk about one area in which safety-related vehicular performance can be quantified there is the fact that the BMW R90S can be made to stop at the rate of 0.9G, while the average passenger car will be straining a brake hose to get 0.6G. Thus, the BMW actually is less likely to become a participant at the scene of an accident when whistling along at 80 mph than is the typical sedan moving at a sedate, lawful 55 mph.

Cycle Test

**"An R90/S is—
heart and soul prepared
to give the sporting rider
all the cooperation
he could ask"**

BMW R90/S

Don't bother relating any of the above to the police or the courts; their first allegiance is to law, not reality. The law assumes that there is absolute equality among passenger vehicles, and conversation about fuel consumption rates and stopping distances will avail you naught. Acknowledged or not, the reality is that the new BMW R90S is comfortably, inspiringly unequal to most other road-going vehicles. Unequal, and unlike even its fellows in the world of motorcycles. Other bikes have twin-cylinder engines, the opposed-piston configuration isn't unique, nor the twin-disc front brake, nor shaft drive, and the offset-axle fork design has been around for years. But BMW's technological shopping list always has been peculiarly its own, and today's R90S is a motorcycle of strongly individual character, with a usefulness that extends well beyond the essentially futile gesture of making the law look silly.

No small part of the BMW's usefulness stems from the fact that it is a comfortable fit for its rider. One of the things you don't see, looking at the standard side-view photo of an R90S, is that the bike really is very narrow. Sure, the cylinders jut out like finned fire plugs, but you don't have to straddle them with your knees. All you have between your legs and feet is a tall, narrow fuel tank and an equally tall, narrow engine/transmission casing. There's only 12 inches separating the footpeg rubbers, which is about five inches closer than the same between-pegs gap on a Kawasaki Z-1. The BMW's fuel tank holds 6.3 gallons—a half-gallon of that being reserve—but the capacity is obtained with height and length, not width.

There's only one rider-contact area on the R90S wider than the average for motorcycles and that's the seat, which is a boon to tender backsides. BMW has made the R90S seat a bit thinner than those on its other models, and used a more dense foam for padding. And the differences seem to zero out, because the R90S impressed us as offering a rider perch every bit as comfortable as the soft-saddle R90/6—which is better than just about anything. Passengers will notice some difference, as there's a fender-clearance groove in the rear of the seat pan that thins the padding at that point. Owners of earlier BMWs should take notice of the lateral-pattern ribbing on the seat surface, which has been introduced in response to complaints that the old smooth-top seats didn't offer enough grip to keep R90S riders from slipping back toward the passenger grab-rail under full-throttle acceleration. This, despite the step in elevation between the rider's and passenger's portions of the seat.

Ride an R90S and you'll appreciate how one could become a backslider, both literally and with respect to speed limits. The engine may have only those two cylinders, but they're biggies, fed from a pair of 38mm Dellorto "pumper" carburetors, and working with about as much compression (9.5:1) as a 90mm-bore engine dare have in a world with fuels not always equal to their octane promise. We

The abbreviated fairing isn't streamlining; it deflects the wind from the rider's body.

BMW's opposed-twin engine layout, with air-cooled cylinders and air-warmed toes.

don't know how much horsepower the R90S engine makes. Bikes with chain driven rear wheels we test on Webco's dynamometer, but the shaft drivers would require special and very expensive couplings. It probably doesn't matter much with regard to the BMW: the drag strip figures say there's a goodly amount of sheer horsepower available, especially as compared with the bike's weight; more important to the realities of highway travel is the *kind* of urge the engine develops. The BMW's power simply is always there. You don't have to downshift two gears before the bike will respond to a big handful of throttle. Kawasaki's Z-1 will do the R90S right in the eye at the drag strip, but the BMW pulls smartly away from the Z-1 in a top-gear 60 mph roll-on. It cranks up a lot of power without having to be spinning lots of revs. Open the throttle, and the R90S moves out in a flurry of emphatic, closely-spaced thuds.

Those thuds can be used to embarrass other Superbike riders; they also can be a source of irritation for the R90S rider. They are very substantial 449cc thuds, and they react into the surrounding metal to produce a strong lateral vibration when the engine is at full throttle, and full thud. This vibration may be minimized by closing the gap between thuds, done by using lots of revs, which gets converted by the BMW's tall overall gearing into speeds that will blow transistors in every police radar unit within a hundred miles. The R90S will do those speeds safely, economically and without seriously taxing its engine or rider, and the bike must be an absolute wonder in the *laissez-faire* atmosphere of autobahn touring in its native land. You can't get away with that sort of mach-number cruising here, except in short, furtive spurts, but it's kind of exciting/intriguing/frustrating (choose one) to know you're on a machine capable of making mincemeat of the usual time-distance equations.

There is, on a more mundane level, another way of getting the BMW R90S to justify its reputation for smoothness. You simply click the transmission into 5th and roll with traffic. At a fudge-factored 60 mph, which is in most areas enough to make the highway patrol's eyes narrow but not provoke action, the BMW needs so little throttle to maintain the pace that its thudding weakens almost to the level of imperceptibility. Cruising like that the engine is virtually idling, far more likely to expire of boredom than mechanical stress. If there's a compensating factor for the rider, it is that he is at least being bored comfortably and can devote full attention to such scenery as may be available without distractions from a jouncing ride or bed-of-nails saddle.

If there's anything the BMW doesn't do, it's jounce. The front fork has ultra-soft springs, a 200mm (7.9-inch) travel, and a miniscule amount of static friction, so the front wheel willingly moves to accommodate pavement ripples that would get transmitted straight into the handlebar of any other bike. BMW's offset-axle fork design has a number of advantages, among them is that fore/aft loads are carried over a greater length of fork slider; also that the slider can be bored straight through, instead of being a blind hole, and its actual internal diameter machined to more precise tolerances.

There's an unusual degree of precision in the rear suspension, too, along with a pair of shocks that provide a 125mm (4.9-inch) travel. The swing arm pivots on tapered-roller bearings, and rear wheel alignment is built-in because the shaft-driven axle doesn't have to slide to allow for chain tension adjustments. So the

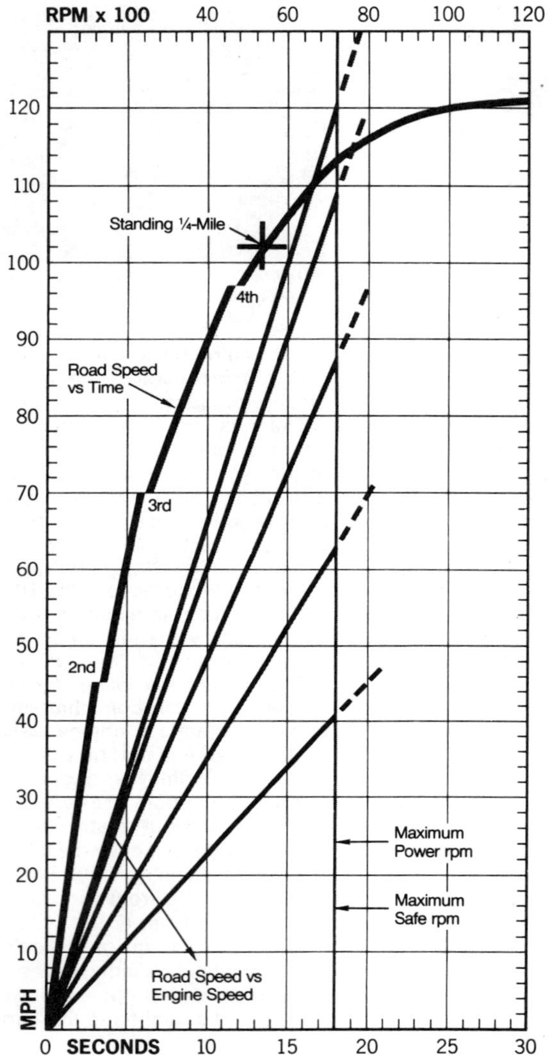

BMW R90/S

Price, suggested retail	$3965
Tire, front	3.25 x 19 H Continental
rear	4.00 x 18 H Continental
Brake, front	1.375 x 10.24 in. x 4 (35 x 260mm x 4)
rear	1.18 x 7.87 in. (30 x 200mm)
Brake swept area	111.7 sq. in. (720.7 sq. cm.)
Specific brake loading	6.0 lbs./sq. in. at test weight
Engine type	Opposed four-stroke twin, OHV
Bore and stroke	3.54 x 2.78 in. (90 x 70.6mm)
Piston displacement	54.8 cu. in. (898cc)
Compression ratio	9.5:1
Carburetion	2; 38mm; PHM Dellorto
Air filtration	Dry paper
Ignition	Battery and coil
Rake/Trail	28°/3.5 in. (90mm)
Mph/1000 rpm, top gear	16.7
Fuel capacity	6.3 gal. (24 liters)
Oil capacity	4.7 pts. (2.25 liters)
Transmission oil capacity	1.7 pts. (0.8 liter)
Electrical power	240 watt alternator
Battery	12V, 2.5 AH
Primary transmission	Helical gears
Secondary transmission	Shaft, spiral-bevel gears, 3.0:1
Gear ratios, overall	(1) 13.20 (2) 8.58 (3) 6.21 (4) 5.01 (5) 4.50
Wheelbase	57.7 in. (146.5cm)
Seat height	32.0 in. (82.0cm)
Ground clearance	6.5 in. (16.5cm)
Curb weight	500 lbs (227.2 kg)
Test weight	665 lbs (320.1 kg)
Instruments	Tachometer, speedometer, tripmeter, ammeter, clock
Standing start ¼-mile	13.18 sec.; 102.04 mph
Average fuel consumption	46 mpg
Speedometer error	30 mph, actual 26.30 60 mph, actual 54.24

BMW gives you soft springing, more wheel travel at both of its ends than anything this side of a new-generation motocrosser, and wheels so precisely aligned that you can tighten the throttle set-screw, lean back and ride for miles without touching the handlebar grips.

Compliance-oriented suspensions and shaft-driven rear wheels have distinct advantages; they also bring with them a couple of penalties. BMW's engineers have integrated the R90S rear hub elements—brake drum, shoes, bevel gears, etc.—into a fairly light assembly, but it still is heavier than its chain-drive counterpart. There are times, as when going quickly over choppy road surfaces, when the BMW's unsprung rear wheel masses get to be a bit more than its springs and shocks can control. And when going quickly you also discover that really hard braking loads the front suspension almost to the compression stops, and that the R90S then acquires a very steep steering-head angle—a hair too steep for steadiness if you're banking into a turn at the same time.

Right here would seem to be the appropriate point to deal with the R90S's handling, which is apt to be the subject of some highly personal judgements. The whole matter can be resolved if everyone understands that the BMW is a sport model, not a racing machine, made for the sporting rider, not racers. It's a question of riding style: racers, especially those of the modern school, sail into corners with brakes applied hard and will use a lot of brake all the way to a turn's apex; the sporting rider tends to do his braking upright, then flop the bike down into whatever attitude conditions require and proceed around under steady throttle until he can see straight road. Assuming approximately equal levels of enthusiasm and skill, there isn't as much difference between the two styles' here-to-there effectiveness as you might suppose. Not unless you're riding an R90S, which is—heart and soul—prepared to give the sporting rider all the cooperation he could ask, but gets all sulky and confused by the racer's overlapping demands.

It is right and proper that the R90S should be suited to the sports rider rather than racers. The latter, with rare exceptions, never have much money left after offering up their bank accounts as sacrifice to their art, and the BMW's price places it exclusively among the fat-wallet crowd, which is far larger than we had imagined. Our first report on the R90S wondered "if the day of the $3000 motorcycle is here." BMW's subsequent sales figures proved that the day had indeed arrived and more recent history, which has brought further price increases, is evidence that there are more motorcyclists among the megabucks bunch than we thought, and that they'll pay whatever is asked for what they want if they think the package justifies the price. BMW's R90S is highly priced . . . and fairly well justified in terms of features and craftsmanship.

One look at the BMW and you know

It's not an illusion: the bike is made left-peg-forward; it's also narrow, which is nice.

it has quality. There is that almost indefinable something we'll call tidiness, and wherever you look pieces fit and sparkle. You get items like the twin brake discs, which have been perforated (a hundred holes in each) to make them lighter and more effective in a streaming rain. Look closely and you'll see that the discs have been cadmium-plated before being surface-ground; BMW doesn't want unsightly rust inside those punched holes. And since the first R90S was introduced the caliper and master-cylinder piston diameters have been increased, because that change improves the brakes' feel. The BMW front brake always worked just great; they thought it should feel more solid. Another feature you could miss is the master-cylinder location, which is under the fuel tank. BMW builds bikes with short handlebars, and they like everything tidy, so instead of attaching a lumpy mechanism at the brake lever pivot they tucked the master-cylinder out of sight, connected it to the lever with a cable—and installed a brake fluid level sensor and warning light.

The brake warning light is up on a panel that carries the speedo and tach, other warning lights for the generator and oil pressure, an amber light that flashes when the turn signals are going, and a green light that's supposed to come on when the transmission is in neutral—but sometimes doesn't. This neutral indicator is in the circuit that energizes the starter solenoid, which means that even if you've found neutral (no mean feat, in the case of our test bike) you can't get any action by thumbing the starter button unless you pull the clutch lever. There's another switch included in the clutch mechanism to override the neutral indicator, so you can leave the bike in first, leap aboard, pull the clutch, hit the starter and make a stylishly fast getaway.

Continued on next page

PHOTOGRAPHY: DALE BOLLER, DAVE HOLEMAN, PAUL R. HALESWORTH, BILL DELANEY

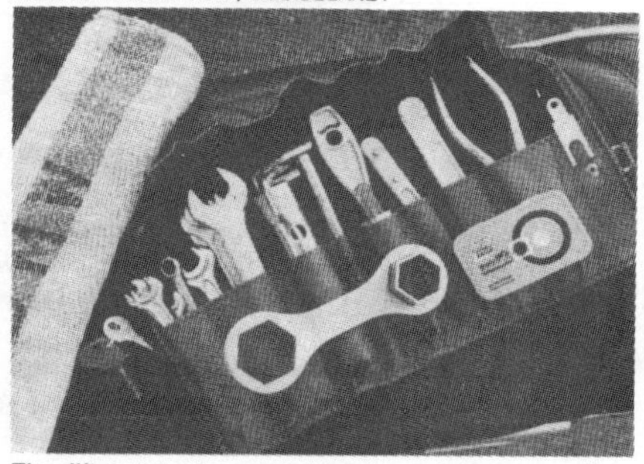

The lift-out tool compartment contains more than two-dozen items, and an official towel.

There are two under-seat storage trays, each with carefully contrived weather sealing.

Instrumentation included an accurate clock, and a grotesquely inaccurate speedometer.

A slotted, knob-controlled device moves the steering damper pivot to give three settings.

Those who buy an R90S should consider leaving it in first-gear anytime it is parked leaned against its side stand. BMW's generally masterful way with machinery seems to have deserted them in the design and execution of the hardware intended to hold the bike upright in its rider's absence. The side stand is a teetery, spring-loaded strut that is just diabolically eager to fold when you want it to hold. Let the R90S roll forward a couple of inches and the next thing you know it has clunked over and is nuzzling the pavement with its left rocker cover. A couple of very near misses persuaded us to make maximum use of the center stand; that was no joy either. Before you can get the center stand deployed it is necessary to grope around under the bike with your toe, like a Romeo trying to play footsy with some lady two chairs away at a crowded dinner party. Once contact is made you work the stand down easily enough—only to find that there's something amiss with contact arcs and fulcrums and you have to haul like mad on the grab-handle above the left side-cover to get the bike's weight up, over and resting securely on the two-legged prop.

Getting the R90S primed to ride is a lot easier than propping it up after you've finished. They've done away with the old side-winder kick-start pedal (which may rightly be considered an anachronism), boosted the starter motor's output by about 20 percent and installed a battery that looks big enough to give the BMW a 50-mile electric-power range after the gasoline is gone. The battery's size reflects undue pessimism on their part. After you've hit the lever that opens the Dellorto carburetors' starting enrichment passages and given the throttle grip a couple of quick twists to prime the cylinders, no more than a touch of the starter button is needed before you get fire and rotation. When the engine is still semi-warm, ignore the choke and just prime the engine with throttle. The BMW R90S carburetors are virtually unique in the motorcycle field in having accelerator pumps, which give each cylinder 0.4cc-squirt of fuel every time you wind the throttle open.

Throttle actuation on the first R90S was mournfully, painfully slow, taking more grip rotation than the human wrist could hope to supply in less than two tries. They've fixed that, and the new throttle mechanism—which still features a pair of bevel gears, and a cam winding in a chain to give progressive action—takes hardly more than a quarter-turn to pull the slides all the way up. This, combined with the excellent response to throttle provided by the Dellorto's accelerator pumps, could have made the bike a bit too sudden for comfort. But the first few degrees of throttle grip rotation give relatively little lift at the slides, due to the camming action of the twist-grip mechanism. It's only when you move the grip farther around that slide movement becomes more rapid. The arrangement is neither as simple nor as cheap as those found on other motorcycles; it does work uncommonly well, and helps explain why the BMW finds so many buyers despite its intimidating price.

The R90S is delivered with other price elevating/justifying features. Like its light aluminum-alloy wheel rims, which follow latest motocross practice in being free of the once-common reinforcing ribs that always collected unsightly muck. And the wheel spokes are made of stainless steel, which is a small item, but important. Other small, important touches are things like the weather sealing that keeps water out of the lift-out tool tray and the odds-and-ends compartment incorporated in the fairing behind the seat. That fairing is, by the way, made of the same impact-resistant molded plastic as the side covers and the small handlebar fairing/windshield. Plastic is a good material in those applications, just as cast aluminum is right for the cleated brake pedal.

BMW has followed long-time habit in making the grab-it end of the ignition key shaped like an instrument knob (black plastic, round and with non-slip serrations) and said key makes an unusually lumpy object in one's pocket ("Are you a BMW owner or are you just glad to see

me?") But you do get a pair of separate fork-lock keys, and the seat latch may be left unlocked, like a car's glove-compartment door. You can get at the tools without killing the engine because you don't have to use the key to spring the lid.

Plan to spend your first weeks of BMW ownership in frequent reaching for the tool kit. Not because the bike will need attention; you simply won't be able to resist showing the kit to your friends. It has everything: combination wrenches all the way from 7mm to 22mm in size, a couple of double-end sockets, a set of feeler gauges for setting valve clearances and spark plug and contact breaker gaps, a set of allen wrenches, the expected pliers and screwdrivers, an odd-looking device used to tighten steering-head bearings and remove fork tube caps, and two very nice tire irons. You will find that the feeler gauges are labeled in German, and we think it only fair to tell you that "Hauptbremszylinder" on the forked feeler means you use it to adjust the free play at the brake master cylinder.

With all these wonders, which tend to make us firmly believe in our theory about the likelihood of a given motorcycle needing roadside repair being in inverse ratio to the quality of its tool kit, perhaps the most impressive item in the tool tray is the hand towel with "BMW" woven right into its fabric. The towel is just the right touch, and all the more impressive after you discover that it is wrapped around a small box of tire-repair patches, and then notice the tire pump clamped handily in place under the bike's seat. You get all these tools, and they mean something, because the BMW owners' manual really is useful. It's the best; better than many workshop manuals we've seen.

This latest BMW R90S is improved over the first version, which we also liked a lot. Much of the bike is all but unchanged—it still has the small clock, which still keeps time accurately—but there are differences and most will not be found looking at the machine in the dealer's showroom. The engine's crankcase castings, for example, have been modified for added stiffness, and the front main bearing, pushrods, cam followers and rocker arms upgraded. There are some detail improvements in the transmission, which did not prevent our test bike from being iffy about engaging 5th gear, and the clutch linkage has been rearranged, in minor fashion, to broaden the engagement point and smooth away some of the clutch grab.

One of the nicest changes is in the front fork damper valving, which has been remetered to provide an extra measure of compression damping, which doesn't keep the R90S from doing a nose-dive under hard braking but does keep it from happening quite so abruptly. Stiffer fork springs are available, and these would make the bike steadier under the pressure of a maximum-effort charge; they'd also make it harsh in ordinary cruising, and harsh just isn't BMW.

We note that some of the BMW R90S models now have Continental tires; others are fitted with Metzlers. Our test bike had Continental tires, and we think their performance, on dry roads at least, is superior. It may well be that the Metzlers are better in the wet, and we'd expect them to be longer wearing because the rubber compound seems to be a bit harder. You pays your money and gets whatever tires happen to come on the bike your dealer has in stock—unless you can argue him into a swap, one way or the other.

With either kind of tire, the R90S is going to be a remarkable time/distance machine. The low, narrow bars hunker you down where the small fairing deflects the worst of the wind-blast away from your torso, and the ride and seating will let you stay with it for hours, as will the huge fuel capacity. Roads stream by under the flashing spokes, the exhaust is a civilized, subdued basso rumble. There's power to spare, reliability unquestioned, and only the omnipresent possibility of red lights twinkling in those big rear-view mirrors to keep you from giving the R90S its head and letting it do what it does best—which is to devour miles in giant bites, in safety and at gas mileages that would make the average sedan owner think he'd surely died and gone to heaven.

BMW R60/6

ONE of our most interesting and informative periods recently was enjoyed with an R60/6 when it was viewed from the angle of an ordinary, budget conscious buyer on the lookout for an economical all rounder. Thinking of BMWs in this manner might seem fanciful, knowing full well the reputation they deservedly enjoy as luxury and somewhat expensive models. £1400 for a 600cc machine might sound exhorbitant, but suppose the bike was judged in a different light, as in fact the one all bikes should be, and most cars already are?

The principle requirements of the desired machine should be listed, including such items as initial cost, home maintenance suitability, likely parts requirements, fuel economy, motorway performance, riding ease, stability, etc, etc. In other words ignoring the normal system of choosing a bike by strict capacity rating and buying, after a series of elimination tests spent honestly evaluating the real performance needs of the buyer, the remaining bike always assuming of course it is a big machine under discussion.

On the assumption that low running costs are required hand in hand with a high cruising speed (or why else buy anything over 250cc?) very few chain driven motorcycles bear comparison, although there are a few exceptions such as the Honda 750F, Yamaha XS650, Laverda GTL; anything over 750cc being in a completely non-utility class by reason of its sheer size and power and all accompanying luxuries. Let's assume the XS650 Yamaha is taken as competition to the R60/6, it being by far the cheapest of the three chain driven bikes at a mere £895. The R60/6 will set you back another £504 on top of that.

On the assumption that regular high speed riding will be practised, and that however carefully maintained no chain — now at £20 each — is going to last more than 4000 miles, an annual mileage of 12000 will cost £60 in chain alone. Add to that lubricant, some spare links, and the usual messy paraphernalia associated with open chain transmission and by the end of five years and 60000 miles, a total of something like £350 will have been spent on the rear transmission alone. It could well be argued that five years is longer than the average owner keeps his machine. Maybe so, but the average owner is not economy conscious, and we are right now, and five years and 60000 miles is a pretty good time to chop a bike in at if it's in good condition.

At the end of that five-year period then, our imaginary big bike budgeter is only £154 down on the chain driven alternative, and he has a machine which even Yamaha would admit is of considerably superior quality all round, and well able to make up that outstanding differential in chain costs at trade-in values.

Obviously other considerations of wear and tear have to be accounted for, but in view of the construction of the R60/6 the likelihood of anything outlasting it is pretty remote.

Top speed was cruising speed, and as top speed proved to be as near 100 mph as makes no difference, then it can be appreciated cruising speed was not far behind and therefore well able to take care of itself among heavier metal, most of which were more powerful and faster, but few of which were able to exploit their advantage, either for legislative or more personal reasons.

After two weeks of constant hammering under the brutal demands of a right fist more used to the power surplus of a truly big engine and therefore one arguably over-riding this one, the machine was returned to BMW. It had bright blue exhaust pipes, was covered in road filth, but was without a damp oil patch to its name anywhere and idled as quietly and evenly as the day it was collected, and if ever there is a test of an engine's quality it's that.

Since the introduction of the 1976 range of machines the improvements within BMWs are all too apparent to anyone trying them for more than a quick nip around the block. As far as this report is concerned, because of the wider margin of "overbuild" already incorporated into the /6 discerning the new mods is extremely difficult. On the R90/S, for instance, its extra power and speed is enough to prove very convincingly that the old, although latterly faint, instability has been driven out completely, and doubtless it requires the general strengthening of engine components, but the likelihood of the 600 ever needing the new mods is rare.

The engine has undergone a revision, although well within existing styling so it appears externally that nothing has happened. In fact the whole thing is a mite wider than previously, but not noticeably so.

The block has a bigger oil sump, or rather a greater area of heat dissipation than previously, although capacity remains the same, thus improving high speed running through heat dispersal via the engine lubricant, a form of cooling increasing numbers of bike makers are looking to these days.

The barrels themselves are wider by two mm than previously, this coming entirely from a thicker base flange designed to increase high speed reliability, and instead of using paper base gaskets, plastic paste is used entirely. Additional oil sealing has been created by the adoption of improved head bolt seals.

BMW R60/6

The camshaft itself is a much thicker one than previously, its bearing surface circumference being increased in diameter from 12 to 20mm, while the front engine cover has been strengthened to suit its camshaft housing.

The biggest noticeable change has taken place in the cylinder head, with new, shorter rockers improving rigidity and with a new striking angle; new aluminium pushrods to reduce valve train reciprocating inertia, and new valve angles all round to improve gas flow and heat transference to the head itself in order to take the load off the exhaust valve particularly.

Small things, but the difference they make, while still very true to BMW traditions, improve things no end, especially at low speeds, at least as far as it is possible to be actually conscious of improvement. Always mechanically quiet, amazingly so for a long pushrod, horizontally opposed twin the new, heavier barrel has incorporated a strengthening rib running across the cooling fins to inhibit ringing, which it does in exemplary fashion. The only mechanical noise issuing from the engine now is a subdued, and oily clickety whirring at idling speeds, inescapable to air cooled pushrod engines.

Top speed might well be higher, although if it is then it is marginal and not measurable as a practical performance increase, but bottom and midrange power is most certainly sweeter. Not much mind, but enough to notice, providing the chance to ride an older model during the same period of time presents itself, and you get the feeling of a sharper power development altogether. Although to claim superior torque would, I feel, be exaggerating a little. More responsive, tighter maybe.

A slightly more powerful starter motor has also been fitted, along with a stronger ring gear around the flywheel to go with it, but I'm blessed if I could tell the difference, although with the bigger capacity battery now fitted undoubtedly the 900cc models benefit, especially during cold weather.

As the engine has been improved, so has the chassis group, although again, it requires the 900 to discern the difference, so adequately was the /6 already catered for, but some measure of the quality of the improvements can be understood when I say, without a shadow of hesitation, that the /S truly is a sportster these days and is no longer the wolf in sheeps clothing it once was.

Most important is the strengthening of the pivoted fork. The root is now heavily gusseted right around the pivot tube and up each leg a little way, and it's made of heavier gauge material all round. To compliment this, the rear suspension units are now carefully paired, although I'm afraid much as I admire the honourable intentions of the factory, the silly infatuation with such practices should by rights be unnecessary with a correctly designed suspension system, for how long a pair of roadster units remains identical is a matter of pure luck.

Up front new, stronger yokes space the tele legs further apart by just under an inch, in order to give clearance to a new set of wider tyres. More importantly the damping system and springing has been changed for the better, and this is probably the most significant performance change I found. They were always good, but perhaps too generously telescopic, for however good the damping was, it could never *quite* contain the eight inches of fork spring action over undulating surfaces. Not that it was ever bad or detracted from high speed security, but the bike has the feeling now of a firmly relaced shoe around your own foot. That's about it. I'm sure you know what I mean.

Both back and front end are firm, in a similar tradition to Moto Guzzi's 750 S3 suspension. Front fork movement has been restricted to around seven inches and, combined with the slightly harder springs, firmer damping puts even the 900s, let alone this little 600, on a par with anything. Because of their low centre of gravity and light weight, line chopping probably exceeds their straight line ability, fine though it is.

Straight line stability was grand. Line chopping utterly safe, but that old bogey I'll swear caused by the unbelievably weak pivoted fork mounting, allowed a slight and quite safe rear end twitch when cornering stresses were at their maximum.

You know, it's awfully difficult, spelling out an accurate description of what is only a slight fault and which barely, if at all, affects the final performance. Despite that the R60 went back with tyre walls more worn than those I've seen on anything else that has passed my way this year. It would appear then, that there is much more to fast riding than stability alone, for the ease with which the twin could be keeled through bend after bend bordered on the miraculous, and I have not met another bike, save the Ducati singles, which could be ridden at such extreme cornering angles at speed, for such long periods without causing either or both, nervous or physical exhaustion.

Light weight, good weight distribution, a set of fine tyres, tenacious damping and gentle power brought it about, and incidentally head grounding was not a problem. It is possible, but without justifiable need on the roads, unless you happen to be on the lookout

for trouble, because travelling fast enough to get that far down, means 10/10s effort, and that's not for me.

Braking was delightful with the sensitive, yet adequately powerful drum. Although criticised by some hard riders as inadequate, it satisfies me gorgeously, if only because it's so reliable, whatever the weather. Why else do so many German buffs discard the discs on the bigger models for racing drums? I'm damned if it can be for power.

All the incidentals that make BMW riding so complete made their presence felt as only good products can. Super big Bosch headlamp, and excellent stands, a sensible tool kit and a finish equal that of a Porsche as far as I could see. A performance that fitted the Porsche iden-

BMW R60/6

tity as well, not Rolls Royce for God's sake as some pretend; that's just plain stupid.

Rotten switchgear though. I hated it when I first saw it and still hate the bloody stuff. Not only is it difficult to operate instinctively, but it was designed to comply to a set of rules made up by lame-brained dopes over on the other side of the Atlantic, trying to apply car legislative techniques to motorcycles, and it should have never have been allowed to cross to this side. In time it will be realised and we will revert to our own more sensible ways, but in the meantime there's this to contend with.

That apart it was a grand bike, almost too good to be considered as an economy model, but then, why not accept that budgeting can be fun as well — if only it's done properly. ●

ULTIMATE MIDDLEWEIGHT

ROAD TEST BMW R60/6

Once upon a time we called 350s middleweights. Then the big bikes started to get bigger and the little bikes got bigger and the bikes which now find themselves in the middle have got 500 or 600 cc engines. By this reckoning the R60/6 is a middleweight — perhaps the ultimate middleweight.

Compared to the 900 BMW the R60 is heavy but in oriental terms its 456lb ready - to - ride puts it fairly and squarely in the middleweight league. Its performance is comparable to the RD350 Yamaha, in fact the Yam would probably out-accelerate it. Price, the stumbling block of all German machines, puts it back in the big bike class. But if you really wanted to convince yourself, once you'd got over the shock of the initial outlay, you could think of the BM

keeping its value while you were having the use of one of the nicest bikes around.

We rode the BMW in January, immediately after the modified /6 range had been announced. I deliberately didn't study the list of modifications to see if I could spot the differences unaided. I couldn't, with the exception that our demonstrator didn't weep oil under the cylinder blocks or around the push-rod housings.

The latest '76 modifications were mentioned last month in our test of the R90S and most of these apply to the R60 as well. The cylinder barrel is wider and its bolts have modified O-rings to prevent oil leaks. The valves are set at a different angle, with shorter rockers and alloy push-rods with new caps, which is said to allow better breathing. If it was done as an attempt to stop the valve adjusters slackening off and clattering, it hasn't worked. Not according to the left hand cylinder on our demo model, at any rate, which was ticking away quite happily after a few hundred miles.

Most of the other mods are concerned with handling. The front fork damping has been changed, the swinging arm is stronger and carries matched dampers but while these changes probably do improve the machine the difference is too slight to be noticeable when compared back over the months since we rode the last BM.

Despite bitterly cold weather the R60 fired up easily every time, unlike earlier models which usually struggled into life on one cylinder and spluttered away until the other finally chipped in. This improvement is probably due to the more powerful starter motor now fitted.

During the test the motor got lumpier at low speeds and caused some vibration at about 3,000rpm as the carburettors got out of synch. As the motor picked up it smoothed out, running cleanly up through the top half of the tacho scale with nothing to be felt from the engine other than its relentless thrust.

The 600 is happy cruising at anything up to about 90mph, easily coping with all kinds of road conditions from motorways to country lanes and city traffic. It's this great adaptability which makes it such an appealing bike. The steering is light and precise enough to make the BM as easy to coax through traffic as a 250 in spite of its size.

Although the motor is less powerful than the R90 the ride and handling are as good so it is easier to use the motor to the full.

By John Robinson

Judging from our fuel consumption figures this is more of a certainty than a probability — on three consecutive tank-to-tank refuellings we got 43.6, 43.5 and 43.5 mpg. This included a spell at the test track, but was still lower than we'd expected and only slightly better than the figures we got for the R90S under equally tough conditions.

When the R60 is used on the open road there is enough all-round performance to make it a very quick motorcycle. Unless hundred-plus cruising and shattering acceleration is demanded, performance is more than adequate and not many machines, whatever their top speed, could keep up the same average speeds with the same margin of safety.

The rest of the BMW range have disc brakes, and they've altered the cylinder sizes to make the single-discs more powerful, but the R60 still relies on the old twin leading shoe drum. Given enough lever pressure this would make the tyre squeal and be controllable enough not to risk locking the wheel but it still isn't as powerful or as effortless as the discs. On the test slopes the front brake wouldn't hold the machine stationary going up the 1 in 4 while the rear one wouldn't stop the machine rolling downhill. As I habitually shift into neutral when a machine stops rolling, this caused some problems going up the slope because the front brake

The rear swinging arm assembly has been strengthened and dampers are matched.

wouldn't hold the bike and sitting with the rear one jammed on didn't leave me with any feet to get it back into gear!

Handling on the road or the track was basically very good. It seemed that the harder the bike was pushed into a corner, the better it would respond. Cruising round more gently made the steering feel lighter and at about 40mph there was a slight yawing as if the handlebar wanted to start flapping around. This model had a pretty good gearshift but it still made the machine twitch on each gear change. It's possible that the 3.25 front tyre was a bit too narrow compared to the 4.00 fitted at the back, and emphasised these skittish tendencies.

While I found that the control layout fitted me and that the switches were easy to locate and use, Bob Goddard didn't get on with them at all. The rocking switches are good if they happen to fall within thumb-span but thumbs obviously vary. One valid criticism is that setting the switches in the lever mounts makes it impossible to adjust one without moving the other.

There was no argument about the lights, though. The QH headlamp is superb with an angled cut-off on dip beam which lets the rider see right along the nearside kerb without dazzling oncoming traffic.

Conclusion: the R60 is an ideal machine for anyone who likes using his bike for long rides and who needs something agile enough for heavy traffic, too. It will do everything asked of it, plus a little bit more. It is this little extra — superb comfort and handling, wide safety margins and the thought which has gone into the design — which distinguishes the BMW. If you think these extras are worth the bike's comparatively high cost, then the BMW is your kind of bike.

For my money it's one of the nicest tourers around and possibly the best middleweight ever, with just one reservation — I haven't ridden the 500 Morini yet.

PERFORMANCE AND SPECIFICATION

TEST CONDITIONS

Dry, ambient temperature 51 deg F, 25 mph headwind

PERFORMANCE

maximum speed		**98 mph**
fuel consumption: hard riding		**43 mpg**
average over test		**48 mpg**
speeds in gears:	**minimum**	**maximum**
first	**7**	**36**
second	**9**	**46**
third	**12**	**76**
fourth	**16**	**92**
fifth	**18**	**106**
		(computed)

standing start quarter mile: 15.9 at 86 mph

ENGINE

599 ccm horizontally opposed OHV twin, bore x stroke 73.5 x 70.6 mm, compression ratio 9.2:1, claimed output 40 hp DIN at 6400 rpm, wet sump lubrication, cb and coil ignition, two 26 mm Bing carburettors, 14V 280 W alternator and 12V 25 A-h battery.

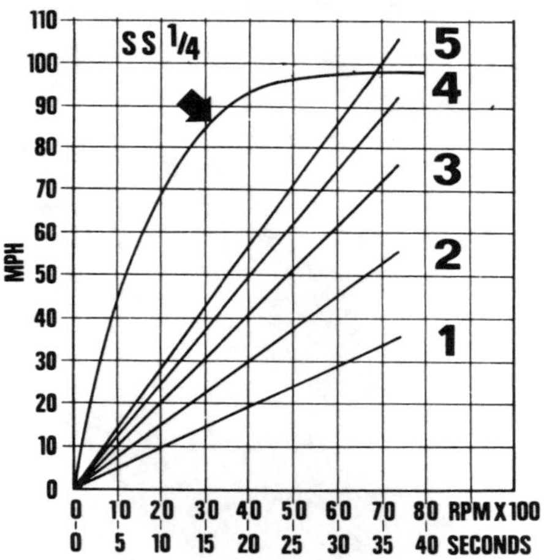

TRANSMISSION

SINGLE plate clutch with diaphragm spring, Cardan shaft and Palloid pattern crown wheel and pinion final drive, 3.36:1 reduction (3.56:1 optional). gearbox ratios: 4.4, 2.86, 2.07, 1.67, 1.50.

CHASSIS

Front tyre, 3.25 x 19, rear tyre 4.00 x 18, telescopic fork front suspension, 8.2 inch travel, swing-arm rear suspension, 3-position dampers, 4.9 inch travel. Wheelbase 58 inch, overall length 85.8 inch, overall width 29.1 inch, seat height 31.9 inch, ground clearance 6.5 inch, dry weight 441 lb, test weight 456 lb, fuel tank 5 gallons, oil capacity 4 pints.

WARRANTY

12 months unlimited mileage, parts and labour.

PARTS PRICES inc VAT

	£
front mudguard	**36.50**
handlebar	**9.45**
speedo cable	**3.56**
exhaust system	**114.80**
pistons/rings	**39.45**
cb points	**3.08**
list price	**1399.00**

HOW IT COMPARES

Model	price	max speed mph	average mpg	ss ¼ mile	dry wt lb
BMW R60/6	1399	98	48	15.9	441
HONDA CB 500T	759	103	50	14.8	425
DUCATI 860	1299	111	42	13.2	452
YAMAHA RD350	580	105	34	14.1	322
HONDA CB400F	699	103	49	15.2	397

BMW R90/6 LIMITED EDITION

Has time changed the big Beemer or left it behind?

Werner, who runs a small sandwich shop near here, thinks BMWs are the greatest thing since knockwurst—but then his father used to work on the BMW assembly line.

A BSA rider pulls up next to the R90/6 at a stoplight and shouts over the roar of his exhaust, "That BeeEm's going to be my next bike!"

Two attendants at a service station eye the big black German bike, and one exclaims, "Wow, it's got a driveshaft."

"Yeah," says the other one, "and a clock!"

BMW motorcycles evoke a lot of positive response from those who look at them. The BMW's reputation for reliability makes it a status symbol within the motorcycling world, and its elegant finish and unique layout even attract the interest of non-bikers.

Motorcycles from the Bayerische Motoren Werke in Munich exhibit a certain point of view about the way a motorcycle should be. Look at it this way: Unlike some of the larger street bike companies, BMW doesn't sell socks; but you can count on your thumbs the number of other companies that offers larger-than-stock gas tanks as original-equipment options.

But BMW is a small company, offering what amounts to one basic design. Each model is fairly similar to the others, with changes in displacement or some external modifications providing differences. Each year brings minor refinements throughout the line, and the company occasionally introduces a variation on the basic theme. The BMW R90 Sport is the best example of variation, but the most recent product of BMW refinement with variation is the subject of this test, the BMW R90/6 Limited Edition.

THE BIKE: The Limited Edition is basically a standard version of the R90/6 in its latest form, equipped with a kit of options . . . at a special price, of course.

As with the other BMW models, the engine is a horizontally-opposed four-stroke twin with a stroke of 70.6mm. The R90's displacement is obtained with a bore of 90mm, and the compression ratio is 9 to 1. A huge, single-piece crankcase houses the one-piece forged crankshaft above the chain-driven camshaft, both of which are placed longitudinally. A spring-loaded tensioner takes up any stretch in the duplex cam chain.

There are two valves per cylinder, and the pushrods that operate them are made out of aluminum instead of steel this year. The external pushrod tubes also serve as passages through which oil can return to the sump to be recirculated by the Eaton trochoid gear-type pump. The pump is driven by the rear of the camshaft, and the front of the cam drives the tachometer cable, ignition point cam, and centrifugal advance mechanism. The front of the crankshaft drives the alternator, which is located under the vented front engine cover along with the points, condensers, advance mechanism, and the charging system's rectifying diodes.

Fuel/air mixture is supplied by a 32mm Bing constant-velocity carb mounted at the rear of each cylinder head. The carbs are the same as found on the 750cc R75/6, and use butterfly chokes to provide an enrichened cold starting mixture. Each carb is given clean air through a tube connected to the airbox at the top rear portion of the engine cases. This airbox and the electric starter, located at the top-front portion of the engine make the engine look tall.

A PCV-type breather allows escaping oil fumes from the crankcase to be passed into the airbox, where they mix with the air/fuel mixture and are burnt before entering the atmosphere. The spent exhaust gases escape through header pipes that exit the front of each cylinder head, and are connected by a crossover tube under the front of the engine.

A large, single-plate dry clutch mounts on an automotive-like flywheel at the rear of the crankshaft. The clutch drives the input shaft of the five-speed transmission, which turns the mainshaft at a reduced speed. The mainshaft drives the layshaft, and the rear end of the layshaft plugs into a universal joint to deliver power directly to the driveshaft inside the right leg of the swingarm. A splined coupling at the rear-wheel end of the driveshaft compensates for variations in the shaft's length caused by the up-and-down movement of the rear wheel.

The final drive ratio is determined by the ratio of the ring and pinion gears in the "rear end." BMW has always approved the fitting of the lower final gearing from the next smaller model. However, the latest generation of BMWs (except for the R90S) now comes with the gearing that was previously standard on the next size smaller machine. Thus, the R90/6 now comes with the 10/32 (3.2 to 1) gearing that used to be standard on the 750 instead of the 11/34 (3.09 to 1) previously used on the R90/6.

The BMW front fork offers more wheel travel than any other machine on the road: 6.25 inches. And the three-way adjustable Boge shocks permit four inches of rear wheel movement. The rear wheel is stopped by a rod-operated, single-leading shoe drum brake, and the front has a single perforated disc that's pinched by a single-action hydraulic caliper mounted behind the fork leg.

The master cylinder for the front brake system is mounted atop the frame backbone under the gas tank, where it is safe from crashes and fiddling fingers. It is linked to the handlebar lever by a cable, and a light on the instrument panel warns when the fluid level is getting low. The alloy wheel rims are laced to the hubs with center-pull spokes, which require less attention than bent-end spokes (which may require frequent tightening).

BMWs come with two stands, as do most road bikes. But unlike almost all other machines, the Beemer's sidestand is spring-loaded for automatic retraction. It is hidden beneath the left cylinder when you're straddling the seat, and may be forgotten—which is why it snaps up as soon as the bike's weight is taken off of it. The bike's centerstand only retracts partially, and must be lifted after the bike is taken off of it.

The most obvious differences between the Limited Edition and the regular R90/6 are the LE's clock and voltmeter, which are mounted in rubber on the upper triple clamp. The merits of clocks on motorcy-

PHOTOGRAPHY: ART FRIEDMAN

In addition to the clock and voltmeter, the Limited Edition has a four-way flasher. The rubber-covered button in front of the steering damper knob operates it.

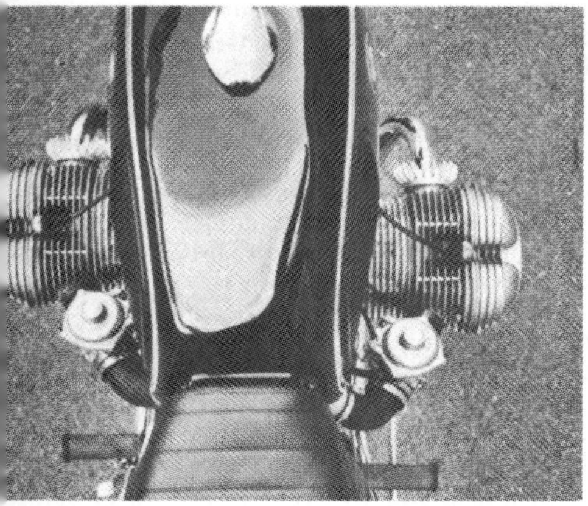

Unless you wear high-topped boots, the carbs' intake tubes will bump your ankles.

There's no slop in the BeeEm's gear-and-chain throttle linkage, but the throttle return pressure helps to give the rider a sore wrist.

cles are debatable, but we find we don't mind them, since it *is* difficult to pull up a tight-fitting jacket sleeve while riding to check your watch. In addition, the LE is the only motorcycle we know of that comes with four-way emergency flashers. They're controlled by a push-button switch clamped to the bars between the instruments. The ignition switch must be turned on before you can start the flashers, but the ignition can be locked and the key removed *after* they're flashing. They may then be left flashing or turned off.

The other special item that's standard on the Limited Edition is stiffer fork springs, designed to be used in conjunction with a touring fairing. BMW is one of the very few companies to realize that a fairing adds 40 pounds or so to the weight of the front of a bike and must be compensated for, so it offers optional springs for use with a fairing on its other bikes. The LE's fork includes these stiffer springs instead of the ones found in the standard R90/6, since the LE is intended to be used with a fairing.

The four-item "kit" adds about $100 to the R90/6 East Coast price of $3395 when purchased as original equipment. The four items would cost about $200 if purchased separately. Our bike was also fitted with a 5.8-gallon fuel tank instead of the standard 4.8-gallon tank, which adds $60 to the price.

ENGINE & GEARBOX: There's no kickstarter, so the R90/6 LE can be started only electrically. It either must be in neutral or have the clutch lever pulled in before the starter will get any juice. The engine is fairly cold-blooded, so the choke must be left on for a minute or two after starting from cold.

The lower rear end gearing has made the overall first gear ratio 14.08 instead of 13.6, so the new R90/6 gets away from a stop with slightly less clutch slippage. That's a good thing, since the clutch's engagement span is relatively brief and ultra-smooth getaways are a little difficult. The clutch is *very* positive—it will not slip—and the engine is very torquey and has a great deal of flywheel inertia, so suddenly dropping the clutch on a lot of rpm will propel you down the street on the rear wheel. Since the torque reaction of the driveshaft tries to make it rotate around the ring gear in the rear end, the BeeEm has a greater tendency to wheelie than would a similar chain-driven bike. Once we remembered to concentrate on keeping the front wheel at least *near* the pavement, the bike sprinted to a standing-start quarter-mile time of 13.46 seconds at 96.9 mph. The R90/6 we tested in June 1974 posted a best standing-start quarter-mile time of 13.49 seconds at 96.8 mph. Obviously, the lower gearing does little to affect the bike's acceleration.

The change in gearing is more noticeable at highway speeds. Out on the open road, you virtually never need to downshift. The bike accelerates with a vengeance when the throttle is opened in fifth gear, and the R90/6 will rush past slower traffic. The bike is one of the *very* best high-gear-only passers we've ever ridden. Fifth gear is more than adequate for climbing—or even accelerating—up long, steep grades, the kind that Harley riders are fond of saying only H-Ds can pull in top gear.

At an indicated 60 mph (actually 62 mph), the big Beemer is humming along at 4200 rpm, just below the 4500 torque peak. Maximum horsepower is developed at 6500, and the tach red line starts at 7100 rpm. The engine pulls willingly from 2000 rpm, although it's smoother and happier above 3000. The constant-velocity carbs keep the engine from balking when the throttle is suddenly opened at low rpm.

The drop in gearing has made the R90/6 nicer around town, too. The gearbox ratios are a little closer together, so it is easier to shift smoothly. We found ourselves shifting with a lurch much less frequently than on BMWs we've tested previously, and we attribute this, at least in part, to the closer gear spacing. And since the gearing is slightly lower, we lugged the engine less often in heavy traffic.

The gearing switch hasn't made any

significant change in gas mileage, however. Our last R90/6 averaged 46.8 miles per gallon, and this one pulled an average of 47.1 miles from each gallon of the specified premium fuel. Incidentally, the 750cc R75/6 we tested last year averaged 45.4 mpg.

As with all BMWs we've ridden, the R90/6's transmission shifted rather heavily, and with a clank. We could shift quickly, but not *very* quickly; and if we tried to make a quick downshift without letting the revs drop low enough, the gearbox crunched and occasionally even resisted going into gear. It sometimes took two or three tries to locate neutral, whether we were rolling or standing still.

The standard BMW torque reaction is present when the throttle is blipped at a stop. The bike responds to the inertia of the flywheel and other engine masses by leaning fractionally to the right. We noticed this reaction only during our first few rides on the bike and then forgot about it. We were never aware of it while the bike was moving.

HANDLING: The Limited Edition's stiffer front fork springs give it a slight advantage over BMWs with standard springs during hard cornering, particularly in smooth corners. The front end doesn't dive quite as much as it does with the standard springs, and this gives the bike a somewhat more stable, confident feel.

However, the stability only goes so far. When we pushed the LE hard in fast turns, it wallowed around considerably; and we were unable to diminish the wallowing by changing the preload settings of the rear shock springs or the position of the hydraulic steering damper. The steering during cornering could not be described as *very* precise, although it is as exact as most machines on the road until it begins to wallow. The big tourer isn't happy when you try to treat it like a road racer.

The cornering ground clearance is slightly improved by the stronger fork springs because the front suspension can't compress quite as much on hard turns. Even though the ground clearance with the standard springs is adequate for a tourer and the LE's springs are intended for use with a fairing, go-fast types might like to use the springs without a fairing. The added stiffness provides better—but still not excellent—stability during intense road-straightening sessions.

The Metzeler tires will let you apply a reasonable amount of cornering pressure on dry pavement, and they work better than most standard-equipment tires on wet pavement. The bike wiggled in rain grooves when we first got it, but by the time the 1200-mile mark had passed, the tires were broken in to the point where only a rare set of rain grooves provoked a wiggle.

Otherwise, the bike was quite stable when rushing down the freeway. It wasn't bothered by sidewinds or the wind wakes of passing trucks. And it didn't wander when it encountered bumps, expansion joints, or potholes. The suspension, even with the stiff springing, keeps the bike well-planted and controllable when negotiating bumpy corners at moderate speeds.

The BMW has 28 degrees of steering head angle and 3.54 inches of front wheel trail. These are moderate figures, aimed at keeping the bike stable without making the steering heavy or too slow. Even at a crawling pace, the steering is light and manageable, and the bike is agile in traffic. However, it doesn't surprise you by roll-steering into turns at low speeds. And it isn't so stable that you have to pull hard on the handlebars or lean a lot to make it turn. Because the "flat" twin provides a low center of gravity, the bike is easy to lean quickly from side to side in an ess-bend.

The bike feels lighter than its 456-pound weight (measured with the fuel tank drained, but with all oils) would lead you to expect, and it isn't as heavy as the massive appearance of the engine would indicate. The LE is light enough to push around with the engine off, and the design of the centerstand makes it easy to get the bike up on the stand.

COMFORT & RIDE: If you ride the R90/6 LE for five hours straight (the approximate range of the 5.8-gallon gas tank), the first things to get sore are your right hand and wrist. Your wrist will hurt mainly because of the large amount of pressure required to hold the throttle open and partially because of the hand grips. Even if you tighten down the friction screw on the throttle drum, you still have to exert some force to open the throttle and keep it from closing. And the throttle grip doesn't have any ribbing, so you have to hold onto it tightly. Combined with the small size and hard rubber of the grips, this will chafe your hand and fatigue your forearm. Furthermore, the bars are bent slightly inward, requiring you to kink your wrists. For long rides at high speeds, we prefer lower, straighter handlebars like the ones found on the R90S.

The seat will keep your hind end comfortable through the first tank of gas if you're riding straight through, and won't bother you until sometime during the second tank. At that point, almost *any* seat would plague you. The Beemer's seat is flat and wide and low enough for our five-foot-two staffer to feel comfortable at a stop. It will accommodate a passenger comfortably on long rides, too.

The footpeg/seat relationship is comfortable, but the pegs are too close to the carburetor intake tubes, so your ankles,

especially your right one, bump the tubes. This is particularly annoying—even painful—if you are wearing shoes that don't cover your ankles. Even though the 5.8-gallon gas tank was very narrow, we usually rode sitting slightly bow-legged so our legs would clear the intake tubes. Carrying a passenger made us sit further forward, so we bumped these tubes even more frequently. When carrying a passenger, we also found that the rider's feet bumped the passenger's feet because the two sets of pegs are very close together.

There is some low-frequency vibration when the engine accelerates below 3000 rpm, but it isn't particularly annoying. Actually, it isn't a vibration as much as it is a transmission of the engine's power pulsations through the chassis. A more annoying, buzzing kind of vibration, mainly in the pegs, starts at about 65–70 mph in fifth gear. This vibration will make your feet tired after several hours on the road—if you care to chance riding that fast. The R90/6 LE wasn't nearly as smooth as the 750 R75/6 we tested.

The LE makes less mechanical noise than any BMW we've ever ridden, and much of the silence is probably due to the new aluminum pushrods. There is a deep, throaty intake/exhaust note when the throttle is opened, but otherwise almost no noise. The sound level was 84.7 decibels as measured by our sound level meter.

The stiffer front fork springs make you slightly more aware of expansion joints, tar strips, and other small bumps, although they are never annoying or fatiguing. If the springs were used with a fairing, the

The engine looks massive and heavy, but the space above the cylinders is occupied mostly by the airbox and the starter.

Virtually all the electrics are located in the engine case or headlight shell, so the area beneath the seat is uniquely uncluttered.

small added bit of stiffness would probably be overcome. And the long-travel fork still absorbs big, deep bumps better than the front suspension on any other street bike.

BMW's handlebar switches recently have undergone a face-lifting, and a kill

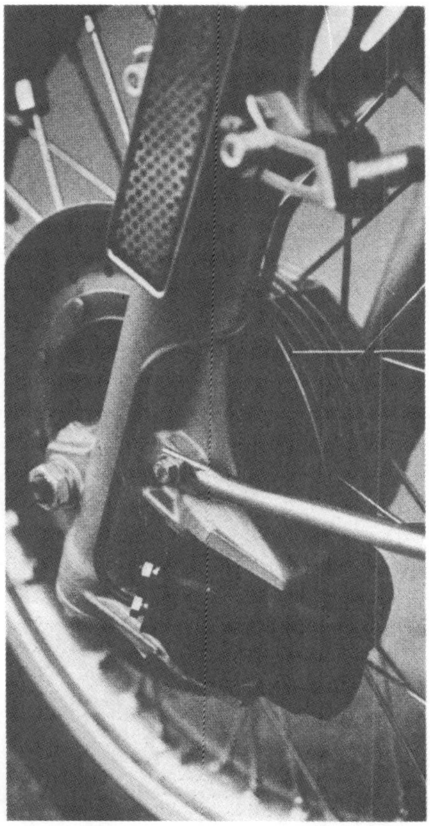

The R90/6 LE needs a more powerful front brake. The holes in the disc are intended to provide better wet-weather stopping.

switch is now in the right cluster. But the new switches leave a lot to be desired. The turn signal switch moves up-and-down instead of left-and-right, and is occasionally confusing. It is located on the right handlebar and gives your already-busy right hand a third function to perform in traffic. The horn button is hard to reach in a hurry, and the horn itself, while louder than many motorcycle horns, is only adequate for around-town driving; it won't discourage anyone who suddenly seems intent upon occupying your lane on the freeway. The high/low-beam switch is also awkward to use. The headlight throws a super-bright beam, but the lighting in the instruments is a little dim. And the speedometer, which is driven from the output side of the transmission, was too pessimistic on our machine.

BRAKING: The R90/6's front disc brake is not particularly powerful, and it's almost impossible to lock the front wheel. This is the main reason for the mediocre panic stops we obtained with the bike. Our best 60 mph stop was 144 feet, 3 inches, and the shortest stop from an actual 30 mph was 34 feet, 5 inches. The feel at the lever is mushy and indefinite, and the front brake is not at all progressive.

The rod linkage to the rear drum brake provides more progressive, surer braking. The brake can be locked if you wish, but it won't surprise you, either by locking unexpectedly or failing to lock when you want it to do so.

The perforations in the front brake disc, intended to provide more positive stopping in the wet (something we had no opportunity to test), make a soft whirring noise when the brake is applied. Neither brake faded or squeaked while we had the motorcycle.

RELIABILITY DURING TEST: We received the R90/6 Limited Edition with 15 miles on it, and recorded 1800 miles during the test. By the time we put 100 miles on the bike, the chrome on the header pipes had turned blue. We also noticed that the center nut for one valve cover was missing, so we had it replaced when we returned the bike for its 600-mile service. A few hundred miles later, the new one was gone, and this time the stud it threads onto was missing, too.

The clutch-operated starter interlock switch (which prevents the starter motor from operating while the transmission is in gear unless the clutch lever is pulled in) worked perfectly when we received the bike. But by the time the odometer showed 800 miles, the switch had failed, and it was necessary to select neutral before starting the engine. The problem is apparently more than a fluke, because we experienced the same failure on the R90S we tested in November of 1974.

The bike required no maintenance work, aside from the 600-mile service. When service *is* necessary, the BMW is laid out so that it can be done easily. Adjusting the valves may be done without any arm contortions, and the points are fairly easy to reach. Getting at the air cleaner element is a little tricky, but you can reach the battery simply by opening the seat and lifting out the tool tray. All the oils are easily changed. Things are made vastly simpler by a good owner's manual and by the complete, high-quality tool kit.

The BMW is the only bike that won't require you to go out and buy a set of tools if you plan to do your own maintenance and repair work. Among other things, the 25-piece tool kit has nine wrenches (including three 10mm wrenches, two open-ends and a box-end), three screwdrivers, five Allen wrenches, a feeler-gauge set, and two tire irons. The tire irons are backed up by a tire pump and a tube patching kit.

SUMMARY: The BMW R90/6 Limited Edition's most important variation from previous models is its lower final drive ratios. The Limited Edition option also includes a clock, a voltmeter, four-way flashers, and stiffer fork springs to make the bike compatible with a touring fairing. Our bike was also equipped with a 5.8-gallon fuel tank, which we would recommend to any touring person contemplating the purchase of a BMW.

The shorter gearing has made the bike

a fantastic fifth-gear passer on the highway and given it a slightly more comfortable performance around town. The bike handles well unless pushed very hard in corners, although the suspension almost makes up for that by offering unsurpassed wheel travel. Although the basic long-distance requirements—a good sitting position, a pleasant ride, and a comfortable seat—are there, the BMW R90/6 LE has some shortcomings that threaten its "optimum tourer" reputation. There is a little vibration, and things like the bars, grips, throttle spring tension, and carb intake tubes need to be sorted out. The braking is just adequate, and certainly not outstanding. We had several small problems, but none caused us any real annoyance or posed a threat to BMW's reputation for reliability.

CONCLUSION: When we first tested the R90/6 back in June of 1974, we remarked that "some of the Japanese multis have closed the gap on BMW in areas where BMW used to stand alone." Since then we've tested a fair number of bikes that reaffirm that statement—bikes like the Suzuki GT750M, the new Honda Super Sport models, and, of course, the Honda GL-1000. And more bikes just around the corner (like the Yamaha 750 and the Honda 750 automatic) promise to make things even more difficult for BMW.

But perhaps the most damning thing for this BMW was that we tested it at the same time as the Yamaha RD400 (found elsewhere in this issue). Our staffers frequently rode one machine just before they got on the other, and we agreed *unanimously* that the RD400, a bike we thought would be a "buzzy little two-stroke" before we rode it, was more comfortable in virtually every way than the R90/6 LE.

We hasten to point out that this doesn't mean the RD400 is a better tourer or is in any way aimed at the same market as the BMW. The Yamaha doesn't have the cruising range, the passing power, or a handful of other features required to put it in the same league as the big Beemer. However, when a sports/commuter bike that needn't count comfort as one of its strong points is found to be more comfortable than an established tourer like the BMW R90/6, you can bet the market is changing quickly.

The R90/6 we tested in June of 1974 wasn't very different from this R90/6 LE, and maybe that is the crux of the problem. Unless some things are changed and improved, the market will bypass BMW, now the standard for touring motorcycles. The R90/6 has quite enough power, economy, range, handling, and reliability to keep it at the head of its class as a tourer. And certainly most of the small things—like the workmanship and the tool kit, for example—will keep its reputation for quality intact. But areas like comfort and brakes must receive attention if the marque is to continue to uphold its traditional position in the sport. ***CG***

SPECIFICATIONS

Engine type	four-stroke
Cylinder arrangement	horizontally-opposed twin
Valve arrangement	overhead valve, pushrod-operated
Bore and stroke	90mm x 70.6mm
Displacement	898.2cc
Compression ratio	9.0:1
Ignition	battery/single point/dual coil
Charging system	12-volt, alternator, rectifier
Carburetion	two 32mm Bing constant-velocity
Air filter	disposable paper element
Lubrication	wet sump, 2.4-qt. (2.3-L) sump capacity
Primary drive	straight-cut gears, 1.5:1 ratio
Clutch	dry, 1 drive plate, 1 driven plate
Starting system	electric only
Transmission	5-speed, left-foot shift
Overall drive ratios	(1) 14.08; (2) 9.15; (3) 6.62; (4) 5.34; (5) 4.80
Pinion gear	10-tooth
Ring gear	32-tooth
Front fork	6.3 in. (160mm) travel
Rear shocks	Boge 3-way adjustable, 4.0 in. (101.6mm) rear wheel travel
Front brake	single-action hydraulic caliper, 10.24-in. (260mm) perforated disc
Rear brake	drum, single-leading shoe, rod-operated
Front tire	3.25S19 Metzeler rib
Rear tire	4.00S18 Metzeler block
Frame	tubular steel, double cradle
Steering head angle	28 degrees from vertical
Front wheel trail	3.54 in. (89.9mm)
Wheelbase	59.3 in. (150.6cm)
Length	86.6 in. (220cm)
Weight	456 lb. (206.8kg)
Weight distribution	45.2% front, 54.8% rear
Ground clearance	7.1 in. (180.3mm) at crankcase sump
Seat height	32.5 in. (825.5mm), unladen
Handlebar width	27.7 in. (703.6mm)
Handlebar grip height	41.2 in. (104.6cm)
Footpeg height	11.6 in. (294.6mm) left, 12.1 in. (307.3mm) right
Instrumentation	speedometer, odometer, tachometer, voltmeter, electric clock
Speedometer error	30 mph indicated, 30 mph actual 60 mph indicated, 62 mph actual
Gas tank	steel, 4.8 gal. (18.2L) standard, 5.8 gal. (22L) optional
Gas consumption	47.1 mpg (20 km/L)
Best ¼-mile acceleration	13.46 sec., 96.6 mph (155.4 kph)
Stopping distance from 30 mph	34 ft., 5 in. (10.5m)
Stopping distance from 60 mph	144 ft., 3 in. (44m)
Sound level per SAE XJ 331a	84.7 db(A)
Suggested retail price	$3555 (as equipped) East and West Coasts

BMW R90/6 LIMITED EDITION

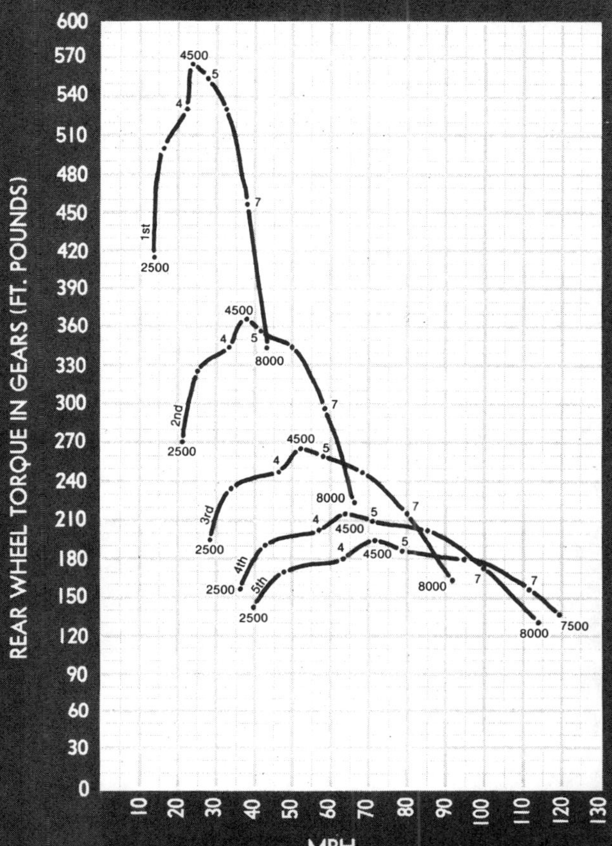

This graph shows the amount of rear wheel torque available at any speed, at any rpm, and in any gear. Maximum acceleration will be obtained by shifting gears at the points where the consecutive lines intersect.

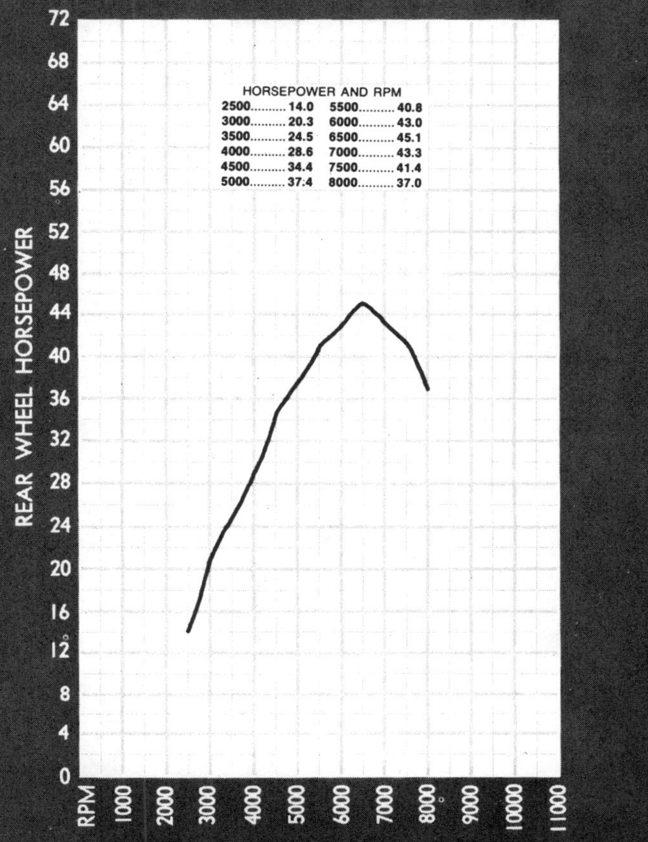

This graph shows the amount of horsepower delivered to the ground as measured by a Patraco MKIII rear wheel dynamometer. These figures may vary from the manufacturer's claims, or from those obtained on a different dynamometer.

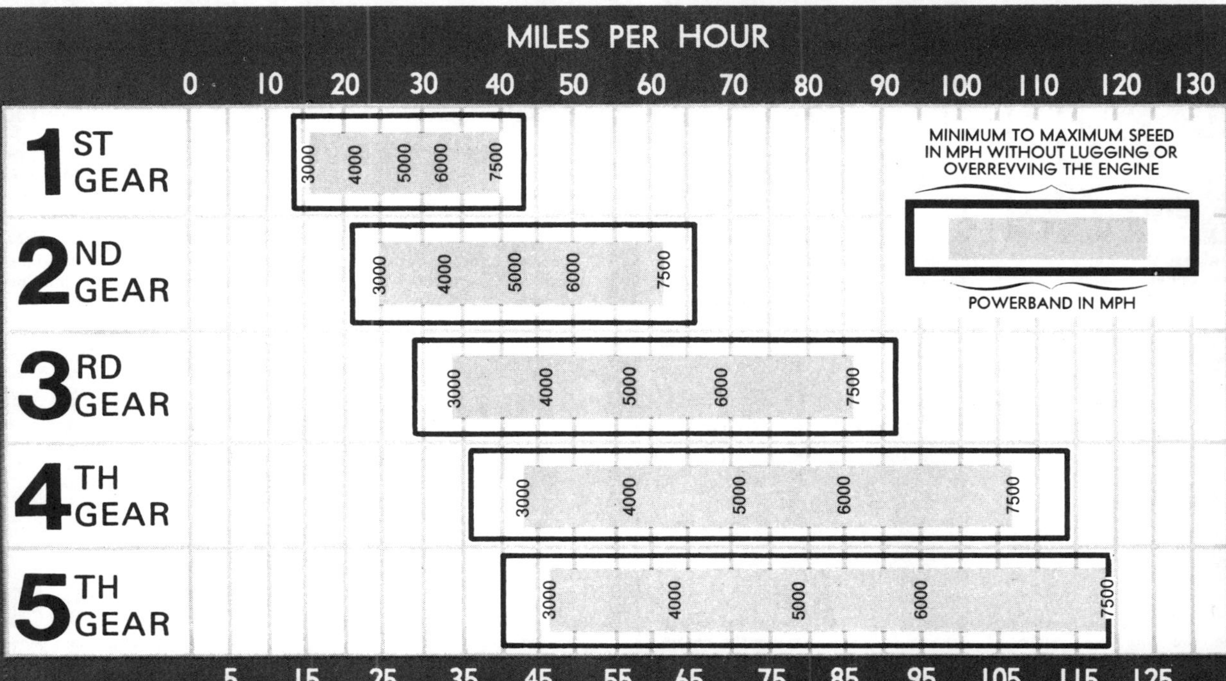

By Brian Lecomber

THE first time I saw a BMW 'R' series engine stripped out on a bench I said to myself: "Coo, what a lovely simple motor to work on. Everything where it ought to be, all nice and uncomplicated . . ."

Which just goes to show how wrong you can be. Or partially wrong, anyway.

In one sense, of course, it is a simple engine. The innards, as you would expect, are well thought out and logical, and the well-equipped home mechanic should be able to knock up what few special tools he needs for a stripdown in half an hour's sawing and drilling.

But after the stripdown, the simplicity ends.

It's not that there's any single facet of the re-building operation

BMW "R" SERIES TWINS

Engine Rebuild series

1. Head and barrel removal is perfectly conventional. Use new gudgeon circlips on re-build.

2. With top ends off, withdraw cam followers. Check for hairline cracks, which cause breakage.

3. Next, take conrods out. Note cap locating pins face towards the front of the motor.

4. Clutch pressure is released evenly by using three studs and nuts in three of the screw holes.

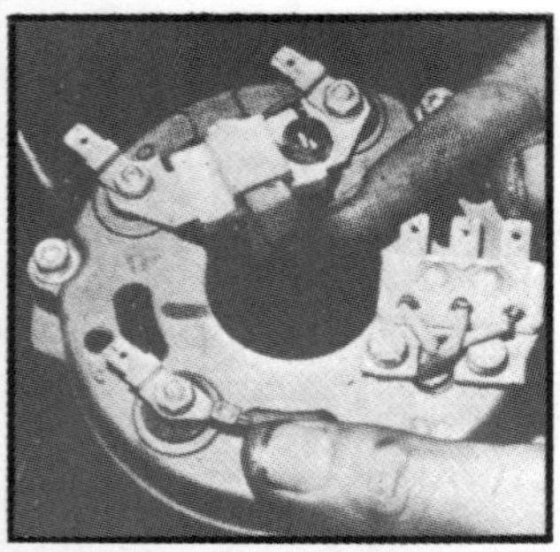

9. Remove the alternator stator, checking condition of the brushes. The rotor requires an extractor . . . .

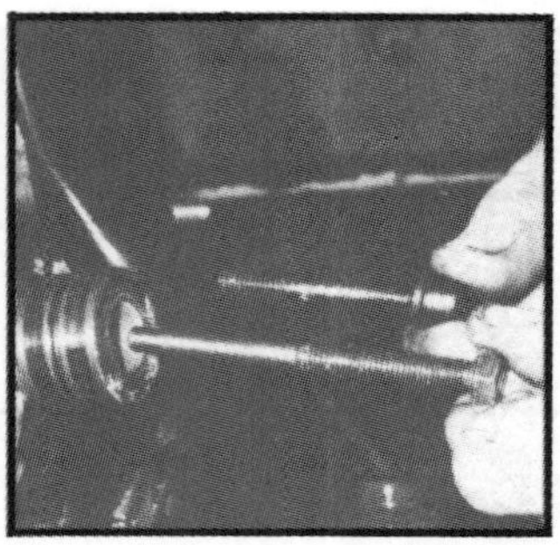

10. . . . . but you can use the centre bolt plus a length of rod equal to extractor's unthreaded length.

11. Rotate camshaft so you can undo the two securing screws. Use Loctite on these on re-build.

12. Remove timing chain tensioner, then use claw extractor to pull crank sprocket off. Cam comes too.

which I can point at with a quavering finger and say: "Hell, nobody could do that on a kitchen table" — but on the other hand, I left the BMW Concessionaires after watching workshop manager Don Clements put an R90S together with the distinct feeling that this engine contains an awful lot of scope for an amateur to make an expensive cock-up. There's that business of drilling out the front main bearing for a start: and then there's the crankshaft end-float performance, and after that the small matter of knocking on the crank sprocket without chewing up the camshaft bearings . . .

Well, you read on and judge for yourself: I'll content myself with saying that I certainly wouldn't advise anybody to tackle a Bee-em as his first engine re-build. Not with the prices of parts what they are nowadays.

The engine comes out of the frame easily enough, and the first part of the stripdown is likewise no problem: you merely undo the six nuts securing each head, remove the rocker assemblies and push-rods, and then heave the heads and barrels off the studs in the usual way. Remember to mark the pistons and barrels Left and Right, and then decide whether you need a re-bore either by the use of micrometers or the time-honoured educated fingernail round the wear-lip. The headwork — decoking, grinding in the valves, etc. — is all perfectly conventional.

The next things to remove are the cam followers — and after you've got them out and labelled them so you know where they go back, it's a good idea to inspect them very carefully for hairline cracks around the edges of the lightening slots. Indeed, the cautious man might even go so far as to have them crack-tested. They have been known to break up, and if they do, then the bits can go everywhere.

The next things off, again after marking them Left and Right, are the conrods. The big-end nuts require a splined socket spanner — not an Allen key — and if you can't get hold of one, try a Ford main dealer.

After that, you can turn your attention to the clutch. Undo three of the six screws and replace them with lengths of studding or long bolts, as shown in the pic, and then run nuts down to the clutch to hold the spring pressure while you remove the remaining three screws. After that, undo the nuts again evenly, so that the clutch plate lifts squarely off the flywheel. Don't try just undoing the screws and letting the plate spring off: for a start it obviously isn't good for it, and secondly you've got to have the studs and nuts to hand for the re-assembly anyway.

The next job is to remove the starter motor, and then pull the flywheel off. This will need an extractor, as shown in the pic, but making same is merely a matter of drilling two holes in a straight bar of metal and then utilising two of the studs and nuts you used for getting the clutch off. Once the flywheel is removed, the plate over the oil pump can be taken off, the guts of the pump removed for cleaning, and then put back so that the tolerances can be measured.

Back at the sharp end of the engine the auto-advance unit is removed, followed by the points-plate and then the alternator stator. The alternator rotor needs an extractor to get it off, but the clever man gets around that by removing the centre bolt, popping a length of rod into the hole to make up the length of the unthreaded portion of the extractor shown in the pic, and then screwing the centre bolt back into the rotor. This makes the bolt itself act as an extractor.

After that, the screws securing the timing cover are removed, and the cover itself tapped off. Underneath it you will find the timing chain, etc.

The first thing is to remove the chain tensioner assembly, and then revolve the engine so that you can undo the two screws securing the camshaft bearing plate. You then take a large claw-extractor, kiss the timing chain goodbye since you will probably be mashing it slightly during your forthcoming labours, and then apply the extractor as shown in order to haul off the crankshaft sprocket and out-rigger bearing. Remember that as the sprocket is slowly withdrawn, so the camshaft must slide out at the same time.

With that lot out on the bench, the back of the job has been broken. The remaining bolts securing the front main bearing housing are removed, and the crankshaft is tapped gently forwards to push the housing out of the case. When that's been done, and the sump has been removed, you have completed the stripdown. And now the fun starts . . .

RE-BUILDING

The first job is to replace the main bearing shells. The rear main is tapped out after heating the crankcase in the oven to 120 degrees C — but before pushing it out, the wise man marks the edge of the bearing housing to show the position of the two oil-holes in the shell, and also the position of the small 'notch' which you'll find cut in the bearing surface. The new bearing is tapped in from the rear, until it is flush with the surface of the case.

To remove the front main bearing shell from its carrier, the first step is to take a small punch and knock out the locating peg (see pic). Then, as with the rear main, mark the position of the

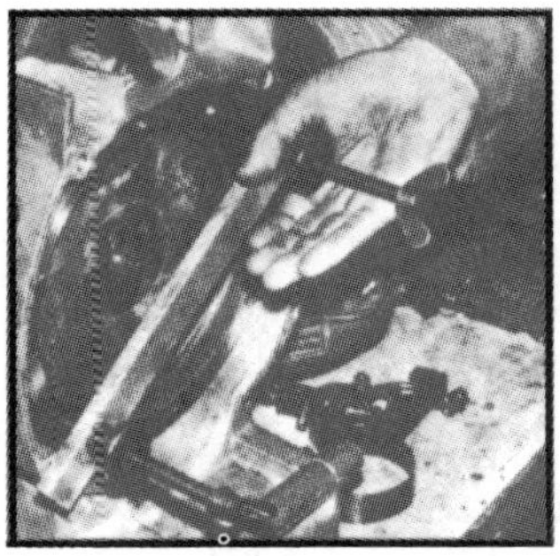

5. Same studs plus a bar can be used to make up a flywheel extractor like this factory job.

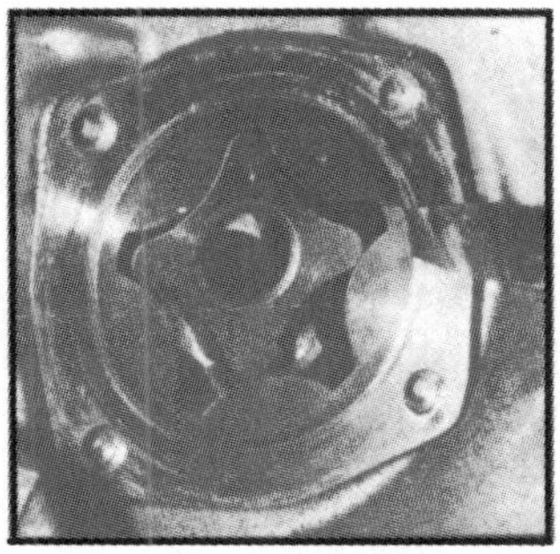

6. With oil pump cover off and pump parts cleaned, check this clearance is no more than .006in.

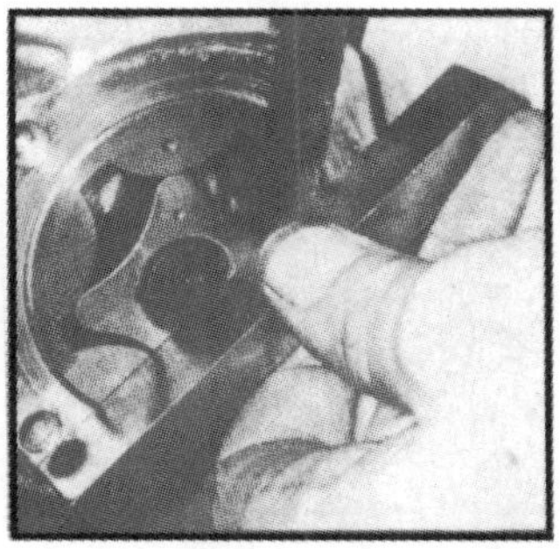

7. Using a straight-edge, check pump end-float. Max. permissible is a tight .002in. for both rotors.

8. Finally, this clearance should be .005—.010in. Dimples on rotors mean this side outwards.

13. Remove oil filter, sump, and sump-strainer. Gaskets should be goo'ed on reassembly.

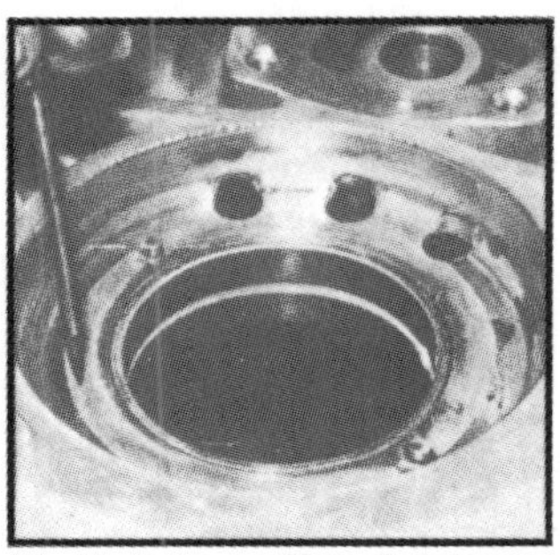

14. Mark the position of rear main shell oil holes, then heat case in oven and tap bearing out.

15. New bearing shell is knocked down until it's flush with the case. Special tool is unnecessary.

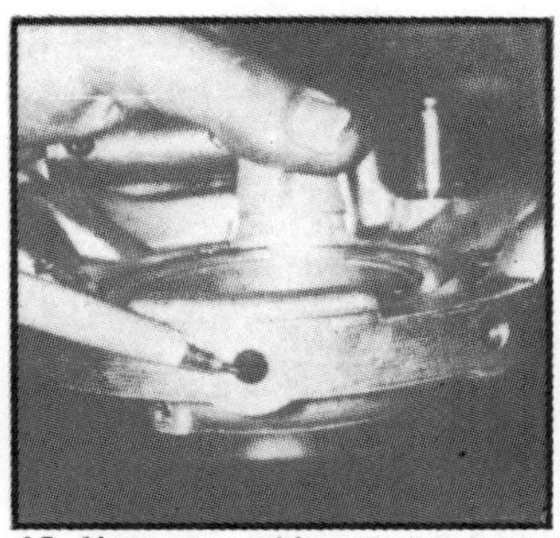

16. Also note position of oil holes in the front bearing shell: They should be north/south.

two oil holes in the centre channel (which should be twelve o'clock and six o'clock), and also the position of the 'notch'. Then heat the housing to 120 C and drive the old bearing out and the new one in.

With the new bearing shell in place, the next step is to run a 3.8 mm drill down the locating peg hole and right through the bearing. You then follow that by running a 4H8 reamer down the hole until you're just making a dent in the shell backing material. This enables you to drive the locating peg back in so that it secures the bearing against spinning, but at the same time sits safely below the level of the bearing surface. Once you have the peg in, burr over the top of the hole slightly so it can't slide out again.

There are now two more holes to be drilled in the bearing. These are the two oil holes at four o'clock and eight o'clock, which issue out on the bearing track itself instead of in the central oil channel. Drilling those is simply a matter of pushing a 3.2 mm drill through the oil holes in the carrier. When you've finished drilling, don't forget to de-burr all the new holes.

Having installed the mains shells, the next step is to set up the crankshaft end float. This is controlled by two thrust washers, one in front of the rear main bearing, and one behind it. These washers won't normally need to be renewed, and if you're using the same crankcase and crankshaft the chances are the old ones will give a satisfactory end-float reading of between .003 and .006 in — but all the same, the end-float does need to be checked.

The thrust washers are colour-coded with tiny blobs of paint (see data chart), and you always use a red one — the thinnest one — inside the crankcase. On the outside — between the crankcase and the flywheel — start off with either another red one or the one that came out. Don't put the rear main oil-seal in on top of the rear thrust washer, because if the end-float ain't right first time you'll have to take it out again, which will invariably screw it up.

With the thrust washers in place, insert the crankshaft and fit the front main bearing carrier roughly, so as to stop the front end of the shaft waving around. (Don't push the bearing carrier right home, as the case needs to be heated up before that's done.). Then bolt on the flywheel, tighten all the bolts to the correct torque, rig up a dial gauge, and check the end float by taking the weight of the whole unit on the nose of the crankshaft and then lifting it by the case. If you've got it right, fine: if you haven't, you'll have to try a different thrust washer under the flywheel. In either case, the flywheel must come off again after the end-float exercise, because you still have the rear main oilseal and the oil pump to fit.

Once the end float has been set up, the rest of the re-build is fairly straightforward. The crankcase must be heated up before the front main bearing carrier is tapped home, and after that you may also find it necessary to

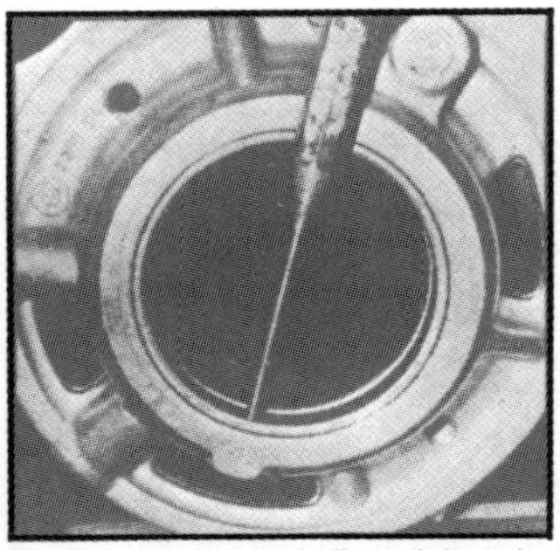

17. Punch out the shell retaining pin as shown, then heat up the bearing carrier, drift the old . . . .

18. . . . . shell out, and drive the new one home until it's flush with the rear surface of the carrier.

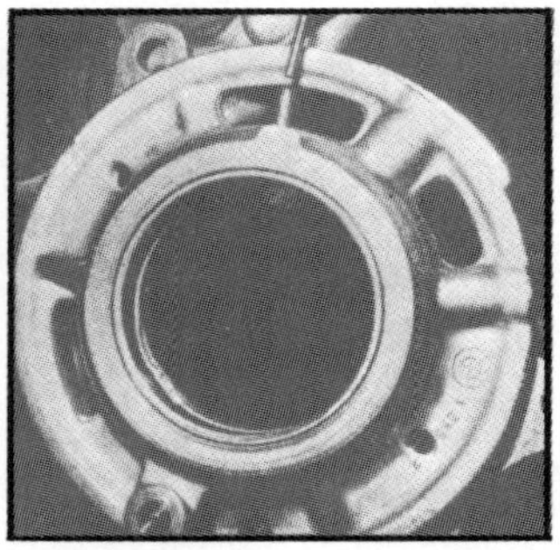

19. Drill and ream locating pin hole (see copy), then tap pin home and peen over hole in carrier.

20. There are two more oil holes which have to be drilled (arrowed). De-burr all the holes afterwards.

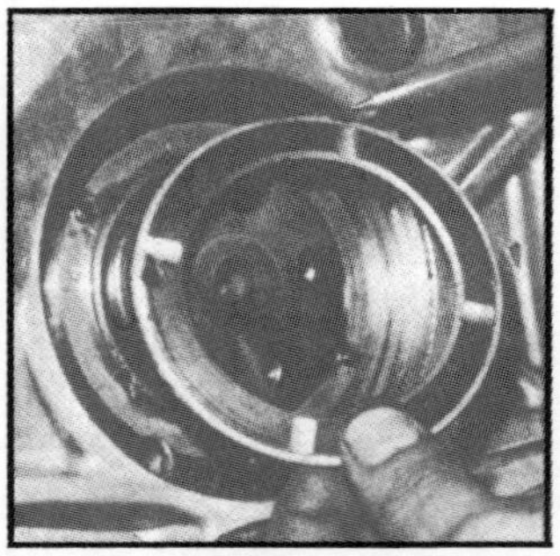

23. Now put either the original rear thrust washer, or another red one, behind the rear main . . . .

24. . . . . and bolt flywheel down to correct torque. Measure end float with dial gauge, and change . . . .

25. . . . . rear washer if necessary. Then heat case and fit front bearing carrier for good.

26. Camshaft end float, measured here, should be .003—.005. Any up-and-down play means new bush.

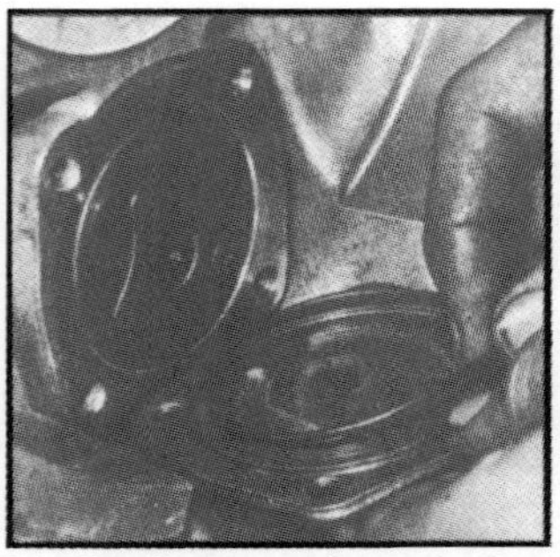

31. Fit oil pump unit, tap home a new rear main oilseal, grease it, and fit the flywheel on.

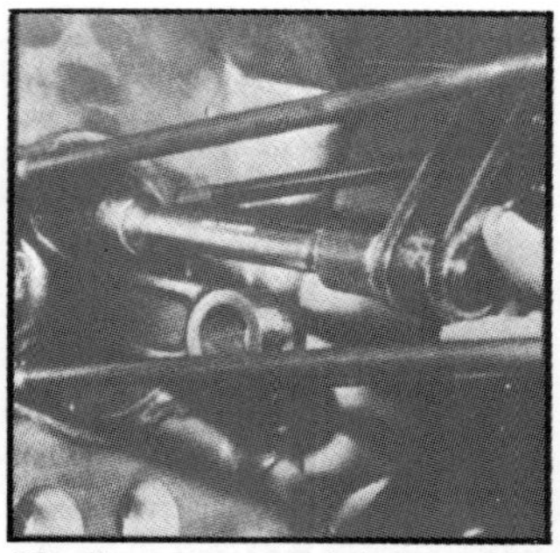

32. Fit new conrod bolts, which require splined socket spanner. Ensure rods are right way round.

33. Pistons are marked *vorn* for front, and have a tiny plus or minus on them for weight grouping.

34. New rings should always be gap-checked. The chrome compression ring goes at the top.

heat up the crankshaft sprocket before tapping it down on the shaft. Make sure you get the valve timing set up (see pics) and the crankshaft sprocket keyway properly lined up with its key before you start thumping that sprocket down, by the way: once it's firmly settled on the shaft you won't be able to turn it, and you can't remove it again without using the claw extractor on your nice new timing chain.

The other thing to watch as you belt the crank sprocket on is the back end of the camshaft. The shaft obviously has to be fed into its location as the sprocket goes home, and if you don't get the rear end properly located in its hole, things are going to get chewed up.

Once the front end's been built up the oil pump can be assembled, the rear main oil seal tapped home and greased, and the flywheel put back for good — not forgetting to make sure that the crank is positioned at TDC and the OT mark on the flywheel is adjacent to the timing hole in the case.

When you come to put the conrods back, there are two things to remember. The first is that the big-end cap locating pins should be towards the front of the engine. Second, you must use new big-end bolts.

The pistons should be checked as per usual — not forgetting the flank clearance of the rings in their grooves — and new rings ought to be pushed into the bores to have their end-gaps checked before fitting. The rings go on the pistons with the chrome ring at the top.

DATA

***Dimensions given in mm., with inch equivalents in brackets.**

		R 60/6	R 75/6	R 90/6 & R 90S
Original bore sizes.	A	73.50 (2.8665)	82.00 (3.1980)	90.00 (3.5100)
(Graded	B	73.51 (2.8668)	82.01 (3.1983)	90.01 (3.5103)
bores).	C	73.52 (2.8672)	82.02 (3.1987)	90.02 (3.5107)

Oversizes available: +0.50 (0.019) and +1.00 (0.039) on all grades.

Min. piston clearance: .035 (.0017)

Max. piston clearance (worn piston/bore): 0.12 (.0046)

Min. ring gap	**0.30 (0.0117)**
Max. ring gap (top rings)	**0.45 (0.0175)**
Min. ring gap (oil ring)	**0.25 (0.0097)**
Max. ring gap (oil ring)	**0.40 (0.0150)**

All rings marked "Top". Pistons marked + or − for weight grading, which mustn't be mixed and "Vorn" for front.

Oil pressure relief valve spring length: 68 (2.65)

Crankshaft end float: 0.08 (0.0031) to 0.15 (0.0059). Thrust washer thickness:

red	**2.53**	**(0.0986)**
blue	**2.578**	**(0.1005)**
green	**2.626**	**(0.1024)**
yellow	**2.673**	**(0.1042)**

Crankpin dia.: 48.00 (1.872).

Main bearing journals: 60.00 (2.34).

Re-grind sizes: −0.25 (−.010), −0.50 (−.020), −0.75 (−.030).

Camshaft end float: 0.08 to 0.12 (.003 to .005).

Valve spring free length: 43.5 (1.696). Fit with green-painted ends to head.

Tappets (set cold): inlet 0.15 (0.006), exhaust 0.20 (0.008).

Points gap: 0.37 (.015).

Timing: 9 degrees BTDC. (S mark visible through timing hole at idle, with strobe.)

Plug gap: 0.6 (.024).

TORQUE SETTINGS

***Given in kg-m, with lb.ft equivalent in brackets.**

Head nuts: 3.9 (28)

Big end bolts: 5.2 (38)

Flywheel bolts: 6.5 (47), (7.5 (54) on 900 models).

Sump bolts: 1.2 (9).

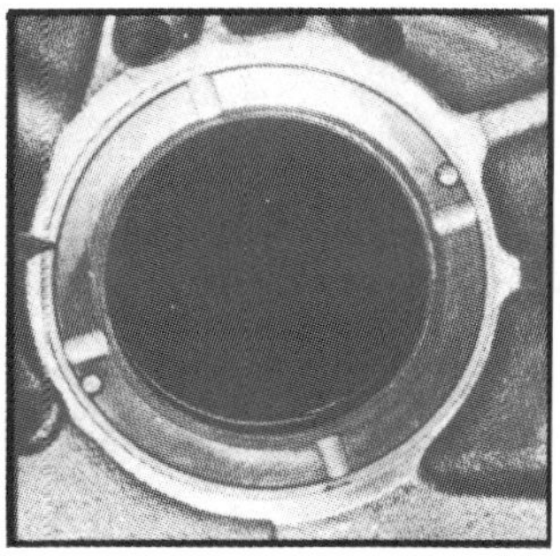

21. Fit a red thrust washer inside the crankcase in front of the rear main bearing. Then put the . . . .

22. . . . . crankshaft in, and enter the front bearing carrier slightly in crankcase to hold shaft steady.

27. Fit new timing chain, with timing marks on sprockets lined up. Align crankshaft sprocket . . . .

28. . . . . carefully with keyway, and tap home on shaft. Watch it, 'cos camshaft enters rear bearing as. . . . .

29. . . . . it goes down along with crank sprocket. Next, tap the outrigger bearing onto shaft.

30. With chain tensioner fitted, tap new oilseals into front cover then tap cover home.

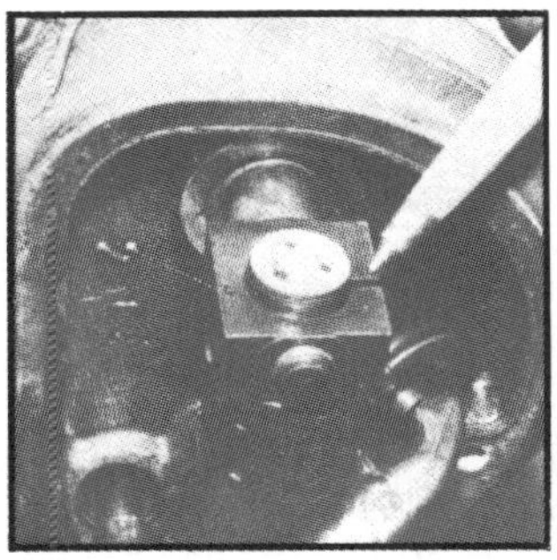

35 Rocker pillars are fitted with their slots facing outwards thus. Slide them on to head-studs . . .

36. . . . . then pinch them together while nuts are nipped up, so as to remove all rocker end-float.

37. When fitting the stator, make sure it *has* entered the housing before tightening the screws.

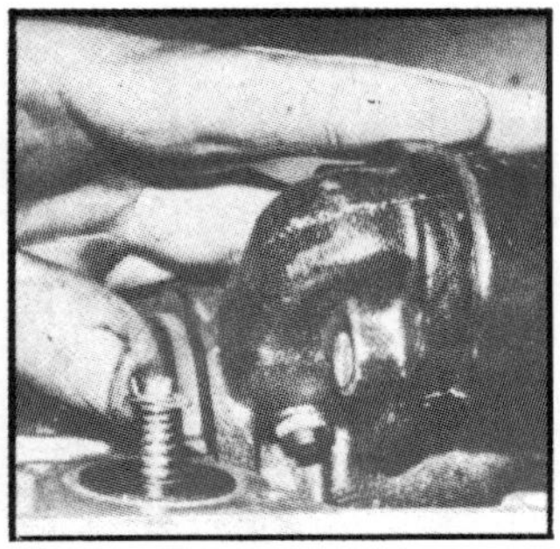

38. Remember to re-fit breather valve. If it chatters, fit circlip in the lower slot.

BMW

Three letters which mean luxury and performance.

Type Usage	Displ.	Engine Type	Bore & Stroke	HP@RPM	Comp. Ratio	Carburetion	Ignition	Gears	Wheel-base	Dry Weight	Aprox. Speed	Aprox. Price

R60/6

With a 600cc horizontally-opposed engine, the R60/6 is the smallest displacement BMW available. It's more than $650 less expensive than its big-bore counterpart, the R90/6. But BMW has not shaved the quality on the R60/6; in fact the most obvious difference is its drum front brake which matches the R60's engine performance. All the familiar BMW features are there: electric starter, enclosed shaft drive, long-travel front fork, super-comfortable seat, self-retracting side stand and even a tachometer. The U.S. model has higher bars than the European version shown here.

Street	599cc	4/S ohv Twin	73.5x70.6mm	46@6600	9.2:1	2, 26mm, Bing	Bat. & Coil	5	57.7"	464 lbs.	100 mph	$2725

R75/6

Much as *Cycle* Magazine's testers liked the R90/6, they preferred the R75/6, contradicting the old principle that bigger is always better. The R75 is smoother than the R90/6; the 750 is geared a bit shorter and runs through the quarter-mile quicker and faster; and the 750 has sharper throttle response and generally feels peppier than the R90/6. The BMW color explosion continues—now the R90/6 and R75/6 will be available in "Havana-Gold" which is a lovely, rich tobacco-leaf brown. You'll be able to get the color on both the standard tank or the 5.8-gallon tank.

Street	745cc	4/S ohv Twin	82x70.6mm	57@6400	9:1	2, 32mm, Bing	Bat. & Coil	5	57.7"	464 lbs.	110 mph	$3125

R90/6

In *Cycle's* Tour Test (August 1975) the R90/6 and Honda's GL-1000 topped the field of eight bikes. Without question, the BMW has the most compliant touring suspension, thanks to its magnificent front fork with 8 inches of travel. The R90/6 is the most comfortable machine on which to spend long hours in the saddle. It has lots of power, the kind that allows you to stack on a fairing, luggage rack, saddlebags and a 150-pound lady-luv without driving the performance down to moped standards. Everything will stop smartly too, courtesy of BMW's large ventilated front disc.

Street	898cc	4/S ohv Twin	90x70.6mm	69@6700	9:1	2, 32mm, Bing	Bat. & Coil	5	57.7"	464 lbs.	115 mph	$3395

R90S

If you looked at production racing at all levels—national and regional—you'd discover that BMWs have been among the most successful brands in the last few years. Consistently you'll find BMWs somewhere in the top three finishing spots. BMW's racing effort continues to make an important statement about all BMWs in general and the R90S in particular: BMWs are performance bikes. The R90S has perforated double-discs upfront; huge 38mm Dell'Orto carburetors; special fairing, tank and seat; and high-speed Metzeler tires. And this year all BMWs have an extended one-year warranty on new bikes. R90S and Rolls Royce are synonymous.

Street	898cc	4/S ohv Twin	90x70.6mm	77@7000	9.5:1	2, 38mm, Bing	Bat. & Coil	5	57.7"	474 lbs.	120 mph	$3965